Read Me First!
A Style Guide for the Computer Industry

Third Edition

Sun Technical Publications

Upper Saddle River, NJ • Boston • Indianapolis • San Francisco
New York • Toronto • Montreal • London • Munich • Paris • Madrid
Capetown • Sydney • Tokyo • Singapore • Mexico City

The publisher offers excellent discounts on this book when ordered in quantity for bulk purchases or special sales, which may include electronic versions and/or custom covers and content particular to your business, training goals, marketing focus, and branding interests. For more information, please contact: U.S. Corporate and Government Sales, (800) 382-3419, corpsales@pearsontechgroup.com.

For sales outside the United States please contact: International Sales, international@pearsoned.com

ISBN-13: 978-0-13-705826-6
ISBN-10: 0-13-705826-8

This product is printed digitally on demand.

Contents

Tables

Examples

Preface

Read Me First! A Style Guide for the Computer Industry provides everything you always wanted to know about documenting computer products, from creating screencasts to documenting web sites, from writing for an international audience to developing a documentation department.

How This Book Is Organized

This book is organized as described in the following paragraphs.

Chapter 1, "Mechanics of Writing," reviews basic punctuation rules and guidelines, plus other general writing rules and conventions.

Chapter 2, "Constructing Text," provides guidelines for tables, cross-references, headings, lists, and other text elements.

Chapter 3, "Writing Style," provides guidelines for writing in a style that facilitates effective communication.

Chapter 4, "Structuring Information," provides information about organizing and presenting your information so readers can find information quickly.

Chapter 5, "Online Writing Style," provides guidelines for writing documentation that is intended primarily for online presentation. Some of these guidelines also apply to online help and web pages.

Chapter 6, "Constructing Links," provides guidelines for using links effectively in online documents.

Chapter 7, "Writing Tasks, Procedures, and Steps," provides guidelines for writing tasks, procedures, and steps in a procedure.

Chapter 8, "Writing for an International Audience," provides guidelines for writing material that is easily understood by readers whose first language is not English and that can be easily translated into other languages.

Chapter 9, "Legal Guidelines," provides guidelines for the proper use of copyrights, trademarks, confidential information, and other legal guidelines.

Chapter 10, "Types of Technical Documents," describes the various parts that make up a manual and lists the order in which they appear. This chapter also covers the various types of typical technical documents.

Chapter 11, "Working With an Editor," explains how writers and editors work together to produce high-quality documents.

Chapter 12, "Working With Illustrations," describes illustration formats, styles, and types, and explains how to work with an illustrator. This chapter also provides guidelines for writing callouts, arranging callouts, and writing captions.

Chapter 13, "Writing Alternative Text for Nontext Elements," describes how to write text equivalents, referred to as *alternative text*, for each graphic element in a document to meet one of the key Section 508 accessibility requirements.

Chapter 14, "Documenting Graphical User Interfaces," explains how to document graphical user interfaces (GUIs).

Chapter 15, "Creating Screencasts," explains how to develop and record screencasts. This chapter also includes terminology guidelines and guidelines for writing narration.

Chapter 16, "Using Wikis for Documentation," provides some guidelines related to the documentation aspects of presenting your information in a wiki format.

Chapter 17, "Glossary Guidelines," explains how to create a glossary for a technical manual.

Chapter 18, "Indexing," explains how to prepare an index for a technical manual. This chapter covers issues such as selecting topics to index, style rules for creating an index, and editing the index.

Appendix A, "Developing a Publications Department," provides information about issues related to a documentation department, including topics such as scheduling, roles and responsibilities, technical review, and printing and production.

Appendix B, "General Term Usage," lists proper usage for terms commonly used in technical documentation. This appendix also shows the correct usage for commonly confused words and terms, and lists terms that you should consider avoiding.

Appendix C, "Typographic Conventions," lists common elements mentioned in typical technical documentation for which typographic conventions are recommended.

Appendix D, "Checklists and Forms," contains sample checklists and forms that you can use at various stages of documentation development, including art tracking, print authorization, and a technical review cover letter.

Appendix E, "Recommended Reading," provides an annotated list of reference books related to the field of technical communication.

Changes for This Edition

This latest edition of *Read Me First!* includes:

- New chapters and guidelines
- Structural changes
- Clarification of existing guidelines
- More examples
- Error fixes and consistency improvements

The new material in this edition includes:

- New chapter about how to write alternative text for nontext elements such as screen captures, multimedia content, illustrations, and diagrams
- New chapter about how to create screencasts, screencast terminology, and guidelines for writing narration
- New chapter about writing for wikis and encouraging wiki collaboration
- Tables for symbol name conventions, for common anthropomorphisms, for common idioms and colloquialisms, and for phrasal verbs to avoid
- Updated and expanded recommended reading list

Major structural changes for this edition include:

- Moved general writing guidelines from the previous version of Chapter 4, "Online Writing Style," to appropriate locations in existing chapters and to a new chapter, "Structuring Information"
- Moved the Typographic Conventions table from Chapter 2 to a separate appendix
- Changed the "Numbers and Numerals" section in Chapter 1 to a table format including examples

Acknowledgments

Read Me First! A Style Guide for the Computer Industry is based substantially on the *Sun Microsystems Editorial Style Guide*. The people who contributed to new content for this edition of *Read Me First!* are Julie Bettis, Janice Gelb, Shubha Girish, Jean McVey, Cathleen Reiher, PJ Schemenaur, and Alysson Troffer. Jeff Gardiner was the driving force behind the release of the new edition. Janice Gelb was the project lead.

The people who contributed to the preparation and review of the fifth edition of the *Sun Microsystems Editorial Style Guide* include Julie Bettis, Jan Daugherty, Paul Davies, John Domenichini, Jeff Gardiner, Janice Gelb, Gloria Geller, Martha Hicks-Courant, Sue Jackson, Susan Katz, Damon Kujawa, Antonia Lewis, Mary Martyak, Jean McVey, Steve Posusta, Cathleen Reiher, PJ Schemenaur, SueAnn Spencer, Alysson Troffer, and Linda Wiesner.

Special thanks go to the members of the Sun Microsystems Editorial Forum, who served as primary reviewers and architects of all versions of the *Sun Microsystems Editorial Style Guide*. Bruce Bartlett was instrumental in developing the basis for the chapter on indexing. Thanks also to Andrea Marra, who had the original concept and was the driving force behind the first version. Finally, thanks to Steven Cogorno for invaluable production assistance.

Note – An earlier version of Chapter 8, "Writing for an International Audience," appeared in the *Sun Microsystems Editorial Style Guide* and subsequently in *Solaris International Developer's Guide* by Bill Tuthill.

CHAPTER 1

Mechanics of Writing

Error-free writing requires more than just using good grammar. You must also use correct mechanics of writing in your documents. The *mechanics of writing* specifies the established conventions for words used in technical documentation. *Grammar* reflects the forms of words and their relationships within a sentence. For instance, if you put an apostrophe in a plural word ("Create two file's"), you have made a mistake in the mechanics of writing, not grammar.

The mechanics of writing guidelines in this chapter work well for computer documentation, but other style guides might suggest different rules that are equally effective. In most cases which rules you follow doesn't matter as long as you are consistent within your document or documentation set. See Chapter 2, "Constructing Text," for options related to the use of text and graphical elements, such as section headings, tables, and cross-references.

This chapter discusses the following topics:

Capitalization

Writers tend to err on the side of too much capitalization. The chief reason to capitalize a word is that the word is *proper*, not because the word has greater status than other words. A *proper noun* identifies a specific member of a class. A *common noun*, on the other hand, denotes either the whole class or any random member of the class. For example, King Henry VIII (a particular member of a class) was a king of England (the class itself).

Answering the following question can help you determine whether a noun is proper. If the answer is yes, the noun is probably a common noun.

> Does an article or other limiting word appear before the noun? Limiting words include "a," "the," "this," "some," and "certain."

Notice the difference between the following sentences:

> Use a text editor to change the information in your file.
> Use Text Editor to change the information in your file.

In the first sentence, the article "a" makes clear that the writer is not pointing to a particular member of the group of text editors. Therefore, "text editor" is a common noun. But in the second sentence, the absence of an article or limiting word helps to clarify that the writer is pointing to only one member of the group. In that case, capitalize the proper noun "Text Editor."

Note – See Chapter 17, "Glossary Guidelines," and Chapter 18, "Indexing," for examples of how to capitalize glossary and index entries, respectively.

What to Capitalize

Capitalize the first letter of the first word in the following items, unless the item begins with a literal command name or other literal computer term that is not capitalized:

- Bulleted or numbered lists
- Figure callouts (see "Writing Callouts for Illustrations" on page 238)
- Text in table cells
- Sentences

  When possible, avoid writing sentences that begin with a literal command name or other literal computer term that is not capitalized.

  > **Incorrect:** `format` enables you to divide the disk into slices.
  > **Correct:** Use the `format` utility to divide the disk into slices.

- A complete sentence following a colon

  > The software saves time: You can now press a single key to accomplish what used to take hours of complex calculations.
  >
  > Select from two options: The Save option stores your changes and the Discard option erases your changes.

  For guidelines on colon usage, see "Colon" on page 41.

Capitalize the first letter of the following items:

- The terms "table," "figure," "example," "appendix," "chapter," "section," "part," "step," "version," and "release" when they are followed by a letter or number

  The font style and capitalization in cross-references might differ because these aspects are determined by your authoring environment.

  Go to Chapter 3.
  See Section 9 in the man pages.
  See the following table.

  Note – Do not capitalize "step" when referring to more than one step. For example, write "See steps 6–8."

- Each term that identifies the name of a key on a keyboard

  Control-A
  Escape key
  the M key
  Ctrl+Shift+Q

- The second element of a hyphenated compound word in a title or heading unless the first element is a prefix

  Installing a Half-Inch Disk Drive
  Configuring the Audio-In Component
  Self-correcting Errors

  For guidelines on the appropriate use of a hyphen, see "Hyphen" on page 47.

Capitalize the following items:

- Proper nouns
- The first letter of the first word of a title or heading, and the first letter of all other words in a title or heading *except* conjunctions, articles, prepositions of fewer than four letters, and the "to" in infinitives

  See "Using the Mouse" on page 11.
  How to Delete Text With the Cut Function Key

  Note – For fourth-level headings that you write as complete sentences, use sentence-style capitalization.

- Figure captions, example captions, table captions, and table column headings, using the same rules as for titles and section headings
- The letters of many abbreviations and acronyms

- The Roman numeral that designates the sequence of a part divider in a manual

  Part III

- Hardware switch names and button names

  Power-On/Off switch
  Standby switch
  Power button

What Not to Capitalize

Do not capitalize the following items:

- The word "page" in a page number description

  Refer to page 45.

- The spelled-out words in most acronyms and abbreviations, even though the words ordinarily appear in a shortened form in capital letters

  field-replaceable unit (FRU)
  direct memory access (DMA)

- The "x" in hexadecimal text, as in "0x8E"
- The "x" in "x86" and "x64
- The "x" in dimensions, as in "12x12 inches"
- Any word for the sole reason of emphasizing it (use italic for emphasis)
- Placeholders that the user must replace with real values, unless one or more of the separate words in the placeholder is usually capitalized in running text, for example, "ID."

  Placeholders can appear in command-line syntax, in running text, or in any other place in the document that indicates a user-supplied value.

- Command and function names and other software elements, even in a title or heading

  Names of commands, functions, and other software elements are case-sensitive and should always be written in the same capitalization pattern.

- The second element of a hyphenated compound word in a title or heading if the first element is a prefix

  Self-correcting Errors
  Installing a Half-Inch Disk Drive
  Configuring the Audio-In Component

- Words in figure callouts other than the first word, proper nouns, abbreviations, or acronyms
- The first word following a colon if the word begins a text fragment

  This button has only one purpose: to shut down the system.

Contractions

When using contractions, follow these guidelines:

- Never use a contraction when you want to emphasize the negative.

  Incorrect: Don't press the Escape key.
  Correct: Do *not* press the Escape key.

- Avoid obscure contractions, nonstandard usage, and regionalisms such as "mustn't," "mightn't," "you'd best," "shan't," "ain't," or "don't" to mean "does not."
- Never create your own contractions.
- Avoid adding "'s" for "is" or "has" to form a contraction (for example, "that's").

  This construction can be confused with possessive constructions.

- Do not use the contraction "it's" as an abbreviation of "it is" because of the potential confusion with "its," the possessive of "it."

  If you must use the pronoun "it" or the possessive "its," make sure that the antecedent is clear. In the following example, the antecedent for "its" is unclear and could be either "thread" or "section."

  Incorrect: At the time of its creation, no thread owns the critical section.
  Correct: At the time that the critical section is created, no thread owns it.

  In the following examples, the antecedent is clear.

  Correct: This chapter describes the Comet Connector and its use in the application.

  Correct: The `N` argument is most efficient when it is a product of small primes.

The following contractions are not usually a problem for translators: "can't," "isn't," and "don't" to mean "do not."

Gerunds and Participles

When you use a gerund or a participle, ensure that the phrase or sentence in which the gerund or the participle is used is unambiguous. A *participle* is based on a verb, ends with "-ing" or "-ed," and functions as a modifier. A *gerund* is also based on a verb and ends with "-ing," but a gerund is used as a noun.

Confusion can arise when a gerund is followed immediately by a noun because the gerund could be misinterpreted as a modifier. For example, the sentence "Moving companies can be a growth opportunity in an economic decline" is ambiguous because you can interpret "moving" in either of the following ways:

The movement of companies can be a growth opportunity in an economic decline.

The moving services industry can be a growth opportunity in an economic decline.

Follow these guidelines when using gerunds and participles:

- Rewrite sentences to avoid gerunds that are immediately followed by nouns.

Tip – In many instances, you can avoid ambiguity by preceding the noun with an article or possessive pronoun.

Incorrect:

Take preventive action to avoid disappointing users.

Correct:

Take preventive action to avoid disappointing your users.

Incorrect:

Disabling network services prevents IP packets from doing any harm to the system.

Correct:

The disabling of network services prevents IP packets from doing any harm to the system.

If you disable network services, the IP packets do not harm the system.

- Rewrite sentences to avoid participles that have ambiguous meanings.

The following sentence is ambiguous because you do not know whether the participle "using" applies to the term "request" or "document editor."

The document editor sends an edit message request using the file name as a parameter for the message.

You can interpret this sentence in either of the following ways:

The document editor sends an edit message request that uses the file name as a parameter for the message.

The document editor uses the file name as a parameter for the message to send the message.

The following sentence is ambiguous because you do not know whether the participle "used" applies to the term "variables" or "semaphores."

Semaphores are almost as powerful as conditional variables used with mutexes.

You can interpret this sentence in either of the following ways:

Semaphores are almost as powerful as conditional variables that are being used with mutexes.

Semaphores are almost as powerful as conditional variables when the semaphores are used with mutexes.

Numbers and Numerals

A *number* is expressed by *Arabic numerals* (1, 2, 3, 4), by *Roman numerals* (I, II or i, ii), or by words. *Cardinal numbers* use words such as "one, two, three." *Ordinal numbers* use words such as "first, second, third."

In computer documentation, you most often use numerals when numbers are discussed in text. The following table explains when to spell out numbers and when to use numerals.

TABLE 1–1 Numbers: When to Spell Out and When to Use Numerals

Numeric Element	Guidelines and Exceptions	Examples
< 0	Use a numeral for any negative number.	– 5 degrees C – 5°C (in tables)
0 through 9	Spell out, even in tables.	three computers, one file
	Exceptions Use a numeral in these cases:	
	▪ The number is part of a measurement.	3 MIPS
	▪ The number is used in an approved standard.	layer 6 of the ISO reference model
	▪ The number is of the same type and appears in the same sentence, paragraph, or bulleted list as a number of 10 or greater.	The menu offers 11 options, but you use only 4 options.
10 through 999,999	Use a numeral.	25 computers
	Exceptions Spell out a number above nine in these cases:	
	▪ The number starts a sentence.	Ten files are required.
	▪ The number is immediately followed by a numeral.	twelve 500,000-byte files
1 million or higher	Use a numeral for the first part of the value, and use words to represent the zeros.	3 million instructions
Approximation	Spell out.	hundreds of applications

TABLE 1–1 Numbers: When to Spell Out and When to Use Numerals *(Continued)*

Numeric Element	Guidelines and Exceptions	Examples
Chapter reference	Use a numeral.	Chapter 4
Decimal	Use a numeral. If the value is less than 1, use a leading 0.	1.25 0.15
Example reference	Use a numeral.	Example 5-2
Figure reference	Use a numeral.	Figure 2-7
Fraction	Spell out, except in tables and units of measurement. For more information, see "Using Fractions" on page 34.	two-thirds of the users ½-inch tape drive
Page reference	Use a numeral.	page 24
Part reference	Use an uppercase Roman numeral.	Part IV
Percentage	Use a numeral.	25 percent 25% (in tables)
Section reference	Use a numeral.	Section 6.2
Step reference	Use a numeral.	Step 4
Table reference	Use a numeral.	Table 3-8
Unit of measure, including bytes and bits	Use a numeral. For more information, see "Units of Measurement" on page 38.	6 pounds 6 lb (in tables) 3.5-inch disk drive 10 inches 10 in. (in tables) 12x12 feet 12x12 ft (in tables) 24 megabytes 24 MB (in tables) 8-bit color

TABLE 1–1 Numbers: When to Spell Out and When to Use Numerals *(Continued)*

Numeric Element	Guidelines and Exceptions	Examples
Unit of time	Spell out from zero through nine.	five minutes
	Use a numeral for 10 or higher.	30 minutes
		24 hours
		30 days
	Exception	
	Use a numeral for a unit of time smaller than one second.	5 milliseconds

Punctuating Numbers and Numerals

Numbers and numerals generally require the same punctuation as words. Punctuating numbers and numerals becomes troublesome, however, when the numbers are compounded. Follow these guidelines:

- Do not hyphenate numbers or numerals when they serve as single modifiers.

  Your file contains 500,000 bytes.

- Hyphenate numbers or numerals in compound modifiers.

  Print the 500,000-byte file.

- Do not use a comma in numerals of four digits.

  1028
  6000

- Use a comma in numerals of more than four digits.

  10,000
  600,000

For more information about appropriate use of numbers and numerals, see "Numbers, Symbols, and Punctuation" on page 175.

Using Fractions

The usage of numerals for fractions depends on the context. Sometimes, spelling out the fraction or using decimals is the preferred form. Follow these guidelines:

- Use numerals for fractions in tables and for units of measurement, but spell out common fractions in running text.

  ½-inch tape drive
  half the users in the test

- Use a space between a numeral and its related fraction.

  8 ½ inches

- If a fraction is used in a compound modifier, insert a hyphen between the fraction and its unit of measurement.

  8 ½-inch width

- Use decimals when decimals are the industry standard.

  3.5-inch diskette

- In a table in which you are using a numeric modifier of a fraction to save space, spell out the modifying numeral to avoid confusion.

  In tables: ten ½-inch tape drives (there are ten drives for ½-inch tape)

  In tables: 10 ½-inch tape drive (the drive is for 10 ½-inch tape)

  Preferred in text: 10 tape drives for ½-inch tape

Pronouns

This section provides the following guidelines for the use of pronouns:

- Avoid the indefinite pronoun or indefinite possessive pronoun, especially at the beginning of a sentence, unless the noun to which the pronoun or possessive pronoun refers is clear.

  A pronoun that forces a reader to search for an antecedent can frustrate or mislead the reader. Pronouns that typically cause this type of confusion include "it," "they," "its," "theirs," "this," "these," "that," and "those."

  Incorrect: It also describes how to install the software.

  Correct: This chapter also describes how to install the software.

  Incorrect: You can use these either individually or together.

  Correct: You can use these two options either individually or together.

Incorrect:

The value in this variable is used to determine when to pause during long display output, such as during a software dump. Its value is reset each time the `ok` prompt is displayed.

Correct:

The value in this variable is used to determine when to pause during long display output, such as during a software dump. The variable's value is reset each time the `ok` prompt is displayed.

- Do not use first-person pronouns.

Incorrect:

We recommend that you install the custom components only on large systems.

Correct:

Install the custom components only on large systems.

Incorrect:

We can write a protocol specification that describes the remote version of `printmessage()`.

Correct:

You can write a protocol specification that describes the remote version of `printmessage()`.

Incorrect:

Let's assume that the user already has an account on the system.

Correct:

Assume that the user already has an account on the system.

Technical Abbreviations, Acronyms, and Units of Measurement

Computer documentation requires extensive use of abbreviations, acronyms, and units of measurement, many of which have become generally accepted "words" in the industry language. As with any word in a sentence, use abbreviations, acronyms, and units of measurement accurately and with consistent meaning in your documents. Rely on industry definitions for these terms. Do not create your own abbreviations or acronyms. Reference books of this type include *The New IEEE Standard Dictionary of Electrical and Electronics Terms*, *IBM Dictionary of Computing*, and *Microsoft Press Computer Dictionary*.

Abbreviations and Acronyms

An *abbreviation* is a shortened form of a word or phrase that is used in place of the entire word or phrase. "CPU" for central processing unit, "Btu" for British thermal unit, and "SGML" for

Standard Generalized Markup Language are examples of abbreviations. An *acronym* is an easily pronounceable word formed from the initial letters or major parts of a compound term. "GUI" for graphical user interface, "pixel" for picture element, and "ROM" for read-only memory are common acronyms.

Basic Guidelines for Abbreviations and Acronyms

When using abbreviations or acronyms, follow these guidelines:

- In most cases, expand the term and enclose its abbreviation or acronym in parentheses the first time the term is used in text. Then, continue using the abbreviation or acronym alone.

    A local area network (LAN) consists of computer systems that can communicate with one another through connecting hardware and software. Your company probably uses a LAN.

- Repeat the spelled-out version at least at the first occurrence in each chapter where the abbreviation or acronym appears.

  In online help and other topic-based online documents, repeat the spelled-out version at the first occurrence in each topic. In this context, a topic is typically self-contained and resides in its own file.

- If you are certain that an abbreviation or acronym is a standard term for your target audience, you do not need to spell it out at any occurrence.

  For example, you do not need to spell out "CD-ROM," "RAM," or "CPU" in a system administration book. However, if you are uncertain, err on the side of caution, and spell out the abbreviation or acronym.

- When writing out the full word or phrase, do not capitalize any letters unless the letters are capitalized as part of a standard or begin a proper noun.

    floating-point unit (FPU)
    Internet Protocol (IP)

- Avoid first usage of an acronym or abbreviation in a chapter title, heading, or caption.

  If using an acronym or abbreviation for the first time in a chapter title, heading, or caption is unavoidable, do not enclose the abbreviation or acronym in parentheses next to the full word or phrase. Use either the full word or phrase or the abbreviation or acronym depending on which item is more familiar to your audience. Then provide the explanatory text or the abbreviation or acronym in the paragraph immediately following the chapter title, heading, or caption.

- Avoid using acronyms and abbreviations in the plural form.
- Do not spell out acronyms and abbreviations that are trademarked terms.
- Do not abbreviate trademarked terms.

- When using an acronym, ensure that its pronunciation is natural and obvious to a reader.

  The acronym "SCSI," for example, is pronounced "scuzzy." A user who does not know that "SCSI" is pronounceable might expect to see "*an* SCSI port," not "*a* SCSI port." In such cases, provide a pronunciation key when you first use the acronym by itself, as in this example:

  > A small computer system interface (SCSI, pronounced "scuzzy") cable connects the disk drive to the SCSI port.

- Do not use Latin abbreviations such as e.g., i.e., vs., op. cit., viz., and etc.

  Latin abbreviations can cause problems for translators as well as comprehension problems for native and nonnative English speakers. Because these abbreviations might be obscure for your audience, avoid using them.

- When you need to conserve space in tables or slides, use more abbreviations or symbols than you otherwise would in text.

  Some common space savers include "no." for "number" and "%" for percentages.

Punctuating Abbreviations and Acronyms

While you usually do not have to add punctuation to abbreviations and acronyms, the following list provides a few exceptions:

- Use periods in abbreviations that look like words.

  > U.S. for United States
  > no. for number

- Use punctuation marks other than a period in abbreviations or acronyms when that punctuation is standard form.

  > I/O for input/output
  > 3-D for three-dimensional

Acronyms and abbreviations in the plural form can potentially cause problems for assistive technologies and for localization. However, if you need to use an acronym or abbreviation in the plural form, follow these guidelines:

- Add an "s" and no apostrophe to form the plural of abbreviations or acronyms that contain no periods.

  > PCs
  > ISVs
  > GUIs

- Add an apostrophe and an "s" to form the plural of abbreviations or acronyms that use internal periods.

  > M.S.'s
  > Ph.D.'s

Units of Measurement

When abbreviating units of measurement, follow these guidelines:

- Do not abbreviate common units of measurement, such as inches, pounds, feet, centimeters, and meters, unless space conservation is an overriding concern.

  You may use abbreviations within tables, for example.

- Do not use the # symbol to indicate "pound" or "number," a single quotation mark (') to indicate "foot," or a double quotation mark (") to indicate "inch."

- Use standard abbreviations for units of measurement with great care.

  For example, the difference between Mb and MB is the difference between a megabit and a megabyte. Avoid this confusion by consistently spelling out a term like "megabyte" or by using the less-abbreviated form, "Mbyte."

- Use the format "hh:mm:ss" when abbreviating time units of measurement.

  hh Denotes the number of complete hours that have passed since midnight (in military time)

  mm Denotes the number of complete minutes since the start of the hour

  ss Denotes the number of complete seconds since the start of the minute

  The following is an example of the elapsed time as displayed in a GUI screen:

  ```
  Elapsed Time
  (hh:mm:ss)
   10:54:09
  ```

- Do not add an "s" for the plural of units of measurement.

  Abbreviations for units of measurement already account for plurals.

  For example, the abbreviations for 1 kilowatt and 10 kilowatts are written the same way: kW.

- Use periods in abbreviations of units of measurement that look like words.

  in. for inch

  oz for ounce, lb for pound (because "oz" and "lb" are not words)

- Leave a space between a numeral and an abbreviation unless the industry standard for a particular unit of measurement does not include a space or unless the abbreviation resembles a word.

  12 mm

  220V, 10A

- When using the abbreviation for degrees Celsius or degrees Fahrenheit, do not add a space before or after the degree sign.

  75°F (23.89°C)

- Include the metric or U.S. equivalent of a unit of measurement when appropriate.

  1 in. (2.54 cm)
  0.45359 kg (1 lb)

Punctuation

This section reviews basic punctuation rules and guidelines for American English. These descriptions of common punctuation marks are presented in alphabetical order.

Note – Traditional punctuation marks have specialized meanings in the context of programming languages. A classic example is that of quotation marks in the C shell or Bourne shell. These shells have specialized, nonintuitive meanings for single quotation marks, double quotation marks, and back quotation marks. Watch for these types of specialized usages in your writing and editing.

Apostrophe

Use an apostrophe in the following situations:

- **In contractions.** Use an apostrophe to replace letters that are omitted in a contraction.

  can't
  isn't

- **In place of numerals.** Use an apostrophe to replace omitted numerals. Use this informal construction sparingly.

  Class of '66
  Technology of the '90s

- **For possessives.** Use an apostrophe to denote the possessive case of a noun.

  Add an apostrophe and an "s" to most indefinite pronouns, singular nouns (including collective nouns), and plural nouns that do not end in "s."

  the manager's responsibilities
  someone's system
  the group's privileges
  people's rights
  women's meeting

  Note that inanimate objects can be possessive, for example, "the file's properties."

To form the possessive of singular nouns ending in "s" or its sound, you often add an apostrophe and an "s."

the mouse's buttons
the bus's capacity

Add only the apostrophe when the addition of an "s" results in awkward pronunciation .

Plirg Systems' employees

In a few cases, however, either is acceptable.

M. Travis's files
M. Travis' files

Add an apostrophe to form the possessive of plural nouns that end in "s."

the Travises' files
the boards' interrupts

Add an apostrophe and an "s" to the last word of a compound to form the possessive of most compound constructions.

each other's files
anyone else's business

The possessive of two or more names depends on ownership. In the first example, ownership is joint. In the second example, ownership is individual.

Malcolm and Mary's files
Malcolm's and Mary's files

- **To form plurals.** Use an apostrophe to form the plurals of most numerals and symbols, lowercase letters, and single uppercase letters.

Use an apostrophe to form the plurals of abbreviations and acronyms that use internal periods.

P's and Q's
#'s
1's
Ph.D.'s

The apostrophe is not necessary, although not incorrect, when you are forming the plural of two or more unitary uppercase letters or numerals.

CPUs
user IDs
operating system of the 1990s

Single lowercase letters and single uppercase letters are awkward in the plural possessive form. Rewrite to avoid this problem.

Colon

The following sections describe the appropriate use of a colon.

When to Use a Colon

Use a colon in the following situations:

- **To introduce a list.** When introducing a list, use a colon if the introduction is clearly anticipatory of the list, especially if the introduction contains phrasing such as "the following" or "as follows."

    Default settings include four secondary groups: `operator`, `devices`, `accounts`, and `networks`.

    The following options are available from the Diagnostics menu:

    - Test Computer
    - Inspect Computer
    - Upgrade Software

  If the introduction is complete in itself, use a period. See "Capitalizing and Punctuating Lists" on page 66 for other guidelines to use when punctuating lists.

  Introductory text that ends in a colon can consist of a complete sentence, a sentence fragment, or a noun phrase. However, do not use list items to grammatically complete the sentence fragment. See "List Introduction Guidelines" on page 63 for more guidelines about constructing list introductions.

  When the introduction to steps in a procedure is a complete sentence, the use of a colon is optional. If numbered steps immediately follow the statement, you can generally use a colon. If numbered steps do not immediately follow the statement, use a period.

    Learn how to send a message by following these steps:
    Follow the steps in this section to send a message.

- **Before explanatory text.** Use a colon to indicate that the initial clause will be further explained or illustrated by information that follows the colon.

  The colon serves as a substitute for phrases such as "in other words," "namely," or "for instance."

  Notice in the next example that the first word following the colon is capitalized. Capitalize the first word of the statement if the statement is a complete sentence. Do not capitalize the first word if the statement is a sentence fragment.

    This software package was doomed from the start: Customer requirements were never defined, and management was not committed to the project.

- **After an introduction.** Use a colon after an introduction to a statement or question.

  Here is the choice: Do you want to save the file or delete it?

  Remember this cardinal rule: Never reboot your system until you have saved all of your files.

- **With the name of a disk drive.** Use a colon after the name of a specific disk drive.

  Insert the CD into drive A: and press Return.

When Not to Use a Colon

Do not use a colon in the following situations:

- **To introduce a figure or a table.**

  Figure 3–2 shows the relationship between servers and clients.
  Table 4–7 lists the features and their corresponding UNIX® commands.
  The following figure shows the parts of the editing window.

- **When referring to screen elements in text.** When a field name, menu option, or any element on the screen is followed by a colon, omit the colon in text.

  The Printers menu (even though the onscreen label is "Printers:")
  The Hosts option (even though the onscreen label is "Hosts:")

- **To introduce headings.**

  Incorrect:

  <Head2>Preinstallation Checklist

  Before you begin the installation, verify several things about your system:

  <Head3>Check the Configuration

  Correct:

  <Head2>Preinstallation Checklist

  Before you begin the installation, verify several things about your system.

  <Head3>Check the Configuration

- **At the end of a procedure heading.**

  Incorrect: To Configure Your System:
  Correct: To Configure Your System

- **In a list that is introduced by "includes" or "are" within a sentence.**

  Incorrect:

  The colors that are used in four-color printing are: cyan, magenta, yellow, and black.

  Correct:

  The base colors that are used in four-color printing are cyan, magenta, yellow, and black.

Comma

The following sections describe the appropriate use of a comma.

When to Use a Comma

Use a comma in the following situations:

- **In a series.** Use commas to separate the items in a series of three or more words, phrases, or clauses.

    Among your hidden files are `.cshrc`, `.defaults`, `.login`, and `.mailrc`.

    Using a comma before the conjunction that joins the last two items in a series prevents confusion regarding whether the last two items in a series are related.

    The following sentence is confusing, as the final job opening could be read as a single field ("advertising and public relations").

    Current job openings include positions in programming, technical writing, advertising and public relations.

    If an independent clause already contains a comma, consider using a list to separate the items in a series.

    Incorrect:

    The window has a menu bar, which lists available menus, a palette, which shows graphics tools, and a working area, where you draw.

    Correct:

    The window contains the following items:

    - Menu bar, which lists available menus
    - Palette, which shows graphics tools
    - Working area, where you draw

- **To separate independent clauses in a sentence.** Use a comma to separate independent clauses that are joined by the coordinating conjunctions "and," "but," "yet," "for," "nor," and "or."

    Place the comma before the conjunction.

    You do not have to back up your files, but doing so is prudent.
    She lost all of her work, yet she still does not back up her files.

- **To separate a subordinate clause or long introductory phrase at the start of a sentence from the main clause.**

    If you have not deleted a marked file, you can restore it.
    Using a text editor, change the last line of the file.

- **After a dependent adverbial clause or prepositional phrase that starts a sentence.**

  By recording transactions and automating billing, the financial software saves time and prevents costly errors.

  In such cases, hosts assume that destinations are not accessible.

  Do not include the comma if the phrase appears in its normal order in the sentence.

  Because this feature automatically updates system files, it saves time.
  This feature saves time because it automatically updates system files.

- **To separate an introductory modifier from the rest of the sentence.**

  Hopefully, he entered the personnel office.
  Confident that she had saved her work, she logged out.

- **With nonrestrictive clauses or phrases.** Use a comma to set off nonrestrictive clauses or phrases.

  The mail icon, which looks like a mailbox, flashes.

  If additional contexts are created, as is usually the case, the upper 64-Kbyte limit will be reached with fewer hosts.

- **With parenthetic text.** Use commas to set off short parenthetic material.

  The software, with its simple interface, decreases input time by 50 percent.

- **In addresses.** Use commas to set off components of an address when the address appears in a sentence or on one line.

  Write to Plirg Systems, North Bay Village, Florida.

- **With appositives.** Use commas instead of dashes to set off a single appositive.

  The monitor, hardware that looks like a television set, has only one function.

- **In dates.** Use commas to separate components of a date.

  Use a set of commas to make the year parenthetical when the day of the month is included. The comma is optional, however, with only two components of the date.

  She was hired on January 1, 1996, as a technical writer.
  She was hired in January 1996.

- **With "for example" and similar expressions.** Use commas to set off expressions such as "for example," "that is," and "namely."

  Type the date in MMDDYY format, for example, 110709.

  Precede such expressions with a comma only for minor breaks in continuity. For major breaks in continuity, divide the sentence into two sentences.

Incorrect:

The database lists existing objects; however, it does not include objects created since the previous session.

Correct:

The database lists existing objects. The database does not include objects created since the previous session.

When Not to Use a Comma

Do not use a comma in the following situations:

- **In a series of adjectives that is used as one modifier.**

  Click the small black button at the top of the window.

- **Between two short independent clauses.**

  Back up your work or you are fired.
  Save your changes and quit the text editor.

- **Before a dependent adverbial clause or prepositional phrase that appears at the end of a sentence.**

  A dependent clause or phrase at the end of a sentence is in its expected order in the sentence.

  Because this feature automatically updates system files, it saves time.
  This feature saves time because it automatically updates system files.

- **Before a restrictive clause or phrase.**

  Do not touch the switch that is lit.
  The files to be replaced are in the `/etc/opt` directory.

Dash (Em Dash)

Using an em dash for explanatory purposes can result in sentences that are difficult for readers to understand because the sentences contain more than one main idea. When possible, divide a sentence in which em dashes are used for explanatory purposes into two sentences.

Incorrect:

After a context is established between two peers—say, a client and a server—messages can be protected before being sent.

Correct:

After a context is established between two peers, messages can be protected before being sent. An example of two peers is a client and a server.

Em dashes are sometimes used before and after an appositive series.

> Three vital pieces of hardware—the keyboard, the system unit, and the monitor—are packed in the largest carton.

Dash (En Dash)

Use an en dash in the following situations:

- **To indicate ranges.** Use an en dash, without surrounding spaces, to indicate a range.

    Refer to pages 16–24.
    Place the machines 12–16 inches apart.

  However, if a book uses chapter-by-chapter page numbering, use the word "to" to indicate a page range.

    Refer to pages 2-15 to 2-19.

- **To indicate negative numbers.** Use an en dash as the minus sign for numbers that are less than zero.

    Do not operate this equipment in temperatures lower than –10°C.

- **In lists.** In a bulleted list, you can use an en dash to separate an introductory word or phrase from its explanation.

  When you use this list format, put a space before and after the en dash. If the text following the introductory word or phrase is extensive, use a period instead of an en dash. For an example, see "Writing Bulleted Lists" on page 67.

    The word processing software includes the following features:

    - **Automatic save** – Saves changes every two minutes
    - **Automatic backup** – Creates a backup file when you exit
    - **Automatic recall** – Tracks the last 20 transactions

- **In table, figure, and example numbers.** Most authoring tools automatically provide the table or figure number in table and figure cross-references.

  If you have to type a table or figure number, use an en dash between the two numbers.

Ellipsis Points

Ellipsis points are made up of three dots that do not contain spaces between them. Avoid the use of ellipsis points except when showing truncated text within a code fragment. Do not include the ellipsis points shown in a menu option when mentioning the option in running text.

Exclamation Point

Do not use exclamation points except where they have some technical significance. For example, the ! operators in programming and scripting languages have technical meaning.

Incorrect: Configure the system manually!
Correct: Configure the system manually.

Hyphen

Because the computer industry has developed unique terminology, the use of hyphens has become troublesome. Computer documents are often littered with unnecessary hyphens. As a general rule, hyphenate a multiword expression that is used as a modifier. Do not hyphenate a multiword expression that is used as a verb or noun.

the check-in procedure	check in the material
the direct-access password	if you have direct access
the end-user application	writing for end users
the look-up table	look up the definition

When to Use a Hyphen

Use a hyphen in the following situations:

- **In compound modifiers.** With some exceptions, use a hyphen to form a compound modifier when the modifier is used before the noun.

  An exception is open compound nouns used as modifiers, as described in "When Not to Use a Hyphen" on page 49.

  Review the context-sensitive help.
  This menu-driven application provides all possible options.

  Use hyphens with numerals in compound modifiers.

  Print the 500,000-byte file.

  Note the difference in meaning between "end-user control" and "end user control." If you do not intend to abolish the user's control, use a hyphen to avoid ambiguity.

  Hyphenate a compound modifier when it appears before a noun. When a modifier appears after a noun, do not hyphenate a compound modifier.

  An easy-to-remember mail alias is a person's first initial and last name.

  A mail alias that is easy to remember is a person's first initial and last name.

- **To prevent ambiguity.** Use a hyphen to clarify ambiguous text.

Ed owns a small-doll shop.	Ed owns a small doll shop.
He recovered the sofa.	He re-covered the sofa.

- **With some prefixes and suffixes.** Use a hyphen in most cases between a prefix or suffix and a root word when the combination results in double letters.

  re-enable
  co-organizer
  shell-like

  When in doubt, use the guidelines in a standard dictionary. For example, the following words do not use the general rule:

  preexisting
  reentry
  unnumbered
  misspell

  Use a hyphen to join numbers and proper nouns or modifiers with the following prefixes. However, these prefixes are usually joined without hyphens to common nouns and modifiers.

  mid
  neo
  non
  pan
  pro
  un

  Almost without exception, hyphens join the following prefixes with the main word of a compound:

  all
  ex
  self

- **With two words that precede and modify a noun as a unit if one of the words is a past or present participle.** A participle functions as an adjective and is formed by the addition of "-ing" (present participle) or "-ed" (past participle).

  file-sharing protocol
  write-protected device
  user-defined functions

- **In fractions.** Use a hyphen to separate the components of a spelled-out fraction.

  The resulting file will occupy nearly one-third of your disk.

- **In key combinations for some product lines.** Unless the platform that you are documenting indicates another style, use a hyphen to join simultaneous keystrokes.

  Control-A
  Ctrl-Shift-Q
  Meta-A

  For more information about punctuation for key combinations, see "Documenting Multiple Keystrokes" on page 87.

- **In variable names.** Use a hyphen to separate words of a variable name that is two or more syllables long except *filename*, *filesystem*, *nodename*, *pathname*, *userID*, *username*, and other variable names that are short and easy to read as one word.

  directory-name
  system-name
  hostname
  mount-options

  Do not use a space or underscore in variable names. Reserve underscores for their designated use in code.

Note – Some authoring environments consider a hyphen to be a line-break character. However, some hyphenated terms should stay on the same line, for example, Control-Q. Talk to your tools support person to find out how to indicate a nonbreaking hyphen.

When Not to Use a Hyphen

Do not use a hyphen in the following situations:

- **For industry-accepted terms.** Do not hyphenate compound words that are generally accepted as single words.

  online
  database
  email

- **To construct nouns.** Do not hyphenate two words that are used as a noun even if those same words are hyphenated when they are used as a compound modifier.

  Writing documentation for end users is different from writing the end-user application.

  If you have direct access, you can use the direct-access password.

- **To construct verbs.** Do not hyphenate two words that are used as a verb even if those same words are hyphenated when they are used as a compound modifier.

  Check in the book only after reading the check-in instructions.
  Look up the value in the look-up table.

- **With open compound nouns used as modifiers.** Do not hyphenate open compound nouns used as modifiers except to avoid ambiguity or to comply with industry standards, such as the terms "cathode-ray tube" or "CD-ROM drive."

  An open compound noun is a combination of separate nouns that are so closely related as to constitute a single concept. When using open compound nouns, do not create noun strings longer than three words.

  data interchange format
  device driver interface
  disk storage device
  domain name address
  file name extension
  file server specifications

  For more information, see "Use Modifiers and Nouns Carefully" on page 172.

- **With a compound modifier (adverb) ending in "ly."** Never hyphenate a compound modifier that includes an adverb that ends in "ly."

  An easily remembered mail alias is a person's first initial and last name.

- **By itself in suspended form.** When you have successive compound adjectives with a common component, do not omit the component and leave the hyphen suspended.

  Incorrect: 8- and 7-bit characters
  Correct: 8-bit and 7-bit characters

- **With numerals as single modifiers.** Do not hyphenate numerals or numbers when they serve as single modifiers.

  The file requires 500,000 bytes of disk space.

- **With some prefixes.** The common prefixes listed below usually do not require a hyphen unless a specific word beginning with the prefix is listed as hyphenated in a standard dictionary, or the unhyphenated word is misleading or puzzling (for example, "recover" and "re-cover").

bi	multi	pre
inter	non	sub
meta	over	un
micro	post	under
mini		

- **To indicate a range.** Use an en dash (with no space before or after it) instead of a hyphen to indicate a range.

  Refer to pages 16–24.
  Place the machines 12–16 inches apart.

  However, if a book uses chapter-by-chapter page numbering, use the word "to" to indicate a page range.

  Refer to pages 2-15 to 2-19.

- **With trademarked terms.** See "Proper Use of Trademarks" on page 188 for exceptions.

Parentheses

Avoid parenthetical statements that distract the reader from the main idea of a sentence. If a sentence contains a parenthetical statement, consider rewriting the sentence as two sentences. If the parenthetical statement is a definition, move the term and text to a glossary and create a cross-reference.

Incorrect:

The configuration file that is created by Create Action is written to *home-directory*/`.dt/type/`*action-name*`.dt`. The *action-name* file (the executable file with the same name as the action) is placed in your home directory.

Correct:

The configuration file that is created by Create Action is written to *home-directory*/`.dt/type/`*action-name*`.dt`. The *action-name* file is placed in your home directory. The *action-name* file is the executable file with the same name as the action.

Note – Do not use "(s)" after nouns to indicate singular or plural. Either use the plural alone or insert the phrase "one or more" before the plural.

When to Use Parentheses

Use parentheses in the following situations:

- **In lists.** Use either two parentheses or one parenthesis to set off letters or numerals that designate items that are listed within a sentence.

  Choose from (a) keyboard entry, (b) mouse entry, and (c) voice entry.
  Choose from a) keyboard entry, b) mouse entry, and c) voice entry.

- **With first occurrences of abbreviations or acronyms in text.** Use parentheses to enclose special keyboard symbols, abbreviations, and acronyms when they first appear in text.

The operating system inserts a tilde (~) when a file name is too long.

The software package tracks maintenance on your heating, ventilating, and air conditioning (HVAC) systems.

- **When providing the metric equivalent of a U.S. measure.**

  3 in. (76.2 mm)

- **To enclose an entire sentence.** Parenthetical sentences occasionally can hinder clear writing by causing the reader to pause. When possible, consider rewriting a parenthetical sentence without the parentheses. Occasionally, you might want to use parentheses to enclose an entire sentence that is relevant to information presented in the paragraph, yet dispensable to the paragraph's meaning. When an entire sentence is enclosed in parentheses, place the final parenthesis after the sentence's final punctuation mark.

  Position the pointer on the top scrollbox and click the left mouse button. (For detailed instructions about scrolling windows, see page 586.)

When Not to Use Parentheses

Do not use parentheses in the following situations:

- **With first occurrences of abbreviations or acronyms in chapter titles, headings, and captions.**

  Do not enclose abbreviations or acronyms or their expansions in parentheses in chapter titles, headings, and captions. Use the abbreviation or acronym or the expanded form of a term in a chapter title, heading, or caption. Do not use both a short and a long form.

  Incorrect: VLAN Tags (VIDs) and Physical Points of Attachment (PPAs)
  Correct: VLAN Tags and Physical Points of Attachment

- **To enclose an entire paragraph.**
- **To indicate singular or plural after nouns in the form of "(s)."**

  Instead, either use the plural alone or insert the phrase "one or more" before the plural.

- **To refer to the general form of methods and constructors in object-oriented programming languages such as C++.**

  Including empty parentheses would imply a particular form of the method. Include the word "method" to distinguish it as a method and not a field.

  Incorrect: The `add()` method enables you to insert items.
  Correct: The `add` method enables you to insert items.

  When referring to a specific form of a method or constructor that has multiple forms, use parentheses and argument types. For example:

  The `ArrayList` class has two add methods: `add(Object)` and `add(int, Object)`. The `add(int, Object)` method adds an item at a specified position in this array list.

Period

Use a period in the following situations:

- **To end a sentence.** Use a period to end a declarative or imperative sentence.

  Computer documentation is always grammatically precise.

- **In file and directory names.** Use a period as part of a file name to separate the file name from a file extension.

  When used in technical terms, a period is called a "dot."

  The procedures are in the `howto.doc` file.
  The `ls -a` command lists `.cshrc` and `.orgrc` among your hidden files.

  In the UNIX operating system, a period also serves as an abbreviation for the current directory.

  To copy a file into the current directory, you would type the following command:

  ```
  cp ~/work/budget .
  ```

- **With abbreviations.** A period is used with some abbreviations, and always with abbreviations that would look like a word otherwise.

  a.m.
  U.S.

- **In lists.** In a bulleted list, you can use a period to separate a summary word or phrase from its explanation. If the text following the introductory word or phrase is brief, use an en dash instead of a period in this list format.

  For examples of both strategies, see "Writing Bulleted Lists" on page 67.

Quotation Marks

Use quotation marks in the following situations:

- **For quotes.** Quotation marks indicate that material was taken verbatim from another source.

  Do not enclose verbatim commands, system messages, file names, and so forth in quotation marks. In some cases, a reader can be misled into thinking that the quotation marks are an integral part of text that is to be typed.

- **For multiple-paragraph quotations.** If the quotation has multiple paragraphs, put an open quotation mark at the beginning of each paragraph and a final quotation mark at the end of the last paragraph.

  If the paragraphs are indented as a block quotation, do not use any quotation marks.

- **Around chapter titles and section headings.** Use quotation marks to enclose titles of chapters and headings of sections in a book.

  “Sending Mail” on page 42 describes how to send an email message.

  To customize the default LDAP attributes, see Chapter 6, “Customizing Advanced Features in Mail,” on page 150.

  Online documents typically do not include quotation marks around chapter titles and section headings when the titles are active links.

- **For emphasis.** Use quotation marks to emphasize a word or phrase when it is used in an uncommon way or when it is the subject of discussion.

  Use the `tee` command to take a “snapshot” of your keystrokes.

  The word “menu” is often used in technical writing, but not the word “restaurant.”

  Because of translation concerns, use words or phrases in an uncommon way only when necessary. In many cases, rewriting the text is preferable to using this technique.

  Incorrect:

  Do not run this process from a test book with the same part number as a “real” book.

  Correct:

  Do not run this process from a test book with the same part number as an existing book.

- **Around single letters.** Use quotation marks to surround single letters.

  The letter “x” denotes the horizontal axis.

No single rule governs the placement of quotation marks that are next to other punctuation marks. Whether the final quotation mark follows or precedes another punctuation mark depends on context, as explained here:

- **With commas and periods.** American punctuation style is to place the final quotation mark *after* commas and periods, no matter how long or short the quoted material is.

  “Yes,” he replied, “the program is written.”

- **With colons and semicolons.** Place the final quotation mark *before* a colon or semicolon unless the colon or semicolon is part of quoted text.

  Remember the cardinal rule for taking a “snapshot”: Use the recommended palette.

- **With question marks.** Place the final quotation mark *after* a question mark when the question is part of the quoted material.

  The system prompts, “Do you want to continue?”
  The user's guide answers the question, “What can I do with this product?”

  Place the final quotation mark *before* a question mark that is not part of the quoted material.

  How do I display a list of files that are “hidden”?

Semicolon

Avoid using semicolons. Semicolons are often misused and are difficult to read online. For conjoined sentences, consider rewriting the text as separate sentences.

> **Incorrect:**
>
> The redirects contain the link-layer address of the new first hop; separate address resolution is not necessary.
>
> **Correct:**
>
> The redirects contain the link-layer address of the new first hop. Separate address resolution is not necessary.

Semicolons are sometimes used to separate short independent clauses joined by conjunctive adverbs such as "however" or "therefore."

> Both methods are acceptable; however, the direct access method is preferred.

For serial semicolons, consider rewriting the text as a vertical list.

> **Incorrect:**
>
> The Reply menu provides the following options: Reply (all), include; Reply, include; Reply (all); and Reply.
>
> **Correct:**
>
> The Reply menu provides the following options:
>
> - Reply (all), include
> - Reply, include
> - Reply (all)
> - Reply

Slash

Use a slash (/) character for fractions, unless your authoring environment provides dedicated fraction symbols.

> 1/2
>
> 3/4

Do not use a slash in running text. Slashes can be confusing for translators due to the multiple meanings of this symbol, which can mean "or" (for example, "and/or"), "and or" (for example, "open/close"), and "divide by" (for example, "36/6").

Square Brackets

Square brackets are not substitutes for parentheses. To preserve their unique service as meaningful signals to your readers, construct sentences in a way that minimizes the grammatical need for square brackets.

Use square brackets in the following situations:

- **Within parenthetic text.** Use square brackets to insert a parenthetic word or phrase into material that is already enclosed by parentheses.

  Placing comments within a menu file often makes sense. (See page 154 of *Advanced Skills*, Revision A [May, 2009] for related information.)

- **In optional command-line entries.** Use square brackets to set off an optional part of a command line.

  `date` [*yymmddhhmm*]

CHAPTER 2

Constructing Text

This chapter provides information about the use of text and graphical elements, such as section headings, tables, and cross-references. Many acceptable ways exist to present these elements. The design, layout, and writing style that you choose will help make your document unique.

The style and writing conventions in this chapter include suggestions that have worked successfully in published computer documents. You might decide to adopt these conventions, or to adapt them to your company's documentation. Whatever convention you choose, make sure that the format is easily recognized and understood by readers. Use the style guidelines consistently throughout your document.

This chapter discusses the following topics:

For information about recommended typographic conventions, see Appendix C, "Typographic Conventions."

Headings

Headings describe the material that follows them. The appropriate placement of a document heading depends on the content and flow of information. For example, several pages of material might fit well in the context of a single first-level heading, while a few sentences might require a fourth-level heading. Use first-level headings for the broadest summaries, and become more specific as you progress to fourth-level headings. Do not go beyond fourth-level headings. If you have additional topics to cover beyond fourth-level headings, restructure the text. Try to have at least two headings at each level.

For more information about writing procedure headings, see "Use Procedure Headings Appropriately" on page 151.

Writing Section Headings

Section headings group topics in a chapter and provide points of reference for a reader. Headings are hierarchical. You must carefully build headings and text in a logical and understandable progression.

When writing headings, follow these guidelines:

- Avoid ambiguous one-word headings such as "Overview."

  Such headings are meaningless when viewed out of context, as in a search results list. Add more context to make headings meaningful.

  Incorrect: Overview
  Correct: Application Server Security Overview

- Make sure that the heading summarizes the specific information that is discussed in a section.

  Incorrect: Introduction to Features
  Correct: Plirg X80 Features

  Incorrect: Front Panel
  Correct: Front Panel Features

- Reflect the reader's perspective rather than your own.

  Incorrect: Authentication
  Correct: Administering System Authentication Options

- Keep headings short but as meaningful as possible.

  Concise headings are easier to scan in text and in the table of contents.

  Incorrect: PlirgPak 2.7 Supported Features
  Correct: PlirgPak 2.7 Features

- Place the most important words first.

  Incorrect: Elements of iPlirg High Availability
  Correct: iPlirg High Availability Features

- Try to use parallel construction when writing headings at the same level.

  Incorrect:

  Restoring the Operating System

  Disaster Recovery With Autochanger

  Recovery With a Stand-Alone Drive

  Correct:

  Operating System Disaster Recovery

  Autochanger Disaster Recovery

  Stand-Alone Drive Disaster Recovery

  If a heading level includes gerunds (for example, "Opening," "Installing"), try to write all other headings in the chapter at that same level using gerunds.

  Incorrect:

  Defining Application Server Instances

  Viewing Server Information

  Application Deployment

  Correct:

  Defining Application Server Instances

  Viewing Server Information

  Deploying an Application

  However, do not write a heading using a gerund merely for the purpose of achieving parallelism, as shown in the following example. Note that the heading "Understanding Management Operations" erroneously implies that the user must understand the operations before attempting to perform them.

  Incorrect:

  Understanding Management Operations

  Creating a Managed Object

  Selecting a Managed Object

  Correct:

  Management Operations

  Creating a Managed Object

  Selecting a Managed Object

- Repeat the subject in the first sentence of the paragraph following a heading, rather than using a pronoun to represent the subject.

  Incorrect:

  <Head2>Remote Digital Loopback Test

  This tests the system's ability to...

  Correct:

  <Head2>Remote Digital Loopback Test

  The remote digital loopback test examines the system's ability to...

  Incorrect:

  <Head2>`cd` Command

  Use this to change directories.

  Correct:

  <Head2>`cd` Command

  Use the `cd` command to change directories.

- Avoid starting headings with an article.
- Do not place literal terms without a corresponding noun in a heading.

  For literal command names or other computer terms, include a noun such as "command," "file," or "script" to clearly identify the term.

  Incorrect: `/etc/uucp/Systems`
  Correct: `/etc/uucp/Systems` File

- Do not enclose abbreviations or acronyms or their expansions in parentheses in headings.

  Instead, use the abbreviation or acronym or the expanded form of a term, depending on which form is more familiar to your audience. Then, provide the explanatory text or the abbreviation or acronym in the paragraph immediately following the heading.

  Incorrect:

  <Head2>Introduction to the PlirgSoft API (Application Programming Interface)

  The PlirgSoft™ API is a universal programming interface for access to the PlirgSoft software. You can use the PlirgSoft API to create, open, modify, and print PlirgSoft documents.

  Correct:

  <Head2>Introduction to the PlirgSoft API

  The PlirgSoft application programming interface (API) is a universal programming interface for access to the PlirgSoft software. You can use the PlirgSoft API to create, open, modify, and print PlirgSoft documents.

- Do not repeat the exact text of higher-level headings in subheadings, or the text of chapter titles in section headings.

- Include some text before the initial first-level heading in a chapter.
- Avoid having consecutive headings without some meaningful intervening text between them.

  Do not merely repeat the text of the heading in this intervening text.
- Some documentation groups add qualifiers to chapter titles and section headings to identify the type of information that is contained within them, such as overviews, tasks, and reference material.

  If you use qualifiers, do so consistently within the documentation set.

  Booting a System (Overview)
  Shutting Down a System (Tasks)

Using Fonts in Headings

The font conventions that you use for text in a heading are the same as the conventions that you use for the text in a paragraph. For punctuation and spaces, use the default font. See Appendix C, "Typographic Conventions," for typographic conventions.

Capitalizing and Punctuating Headings

This guide follows American English punctuation and capitalization guidelines for headings. Follow these guidelines:

- Use no punctuation at the end of a heading except for a question mark or quotation mark when needed.
- Capitalize the first letter of the first word and the first letter of all other words *except* conjunctions, articles, prepositions of fewer than four letters, and the "to" in infinitives.
- Capitalize the second element of a hyphenated compound word unless the first element is a prefix.

  Structure of Table-Based Documents
  Using a Site-Specific Installation Program
  Initiating Self-checking

Numbering Section Headings

Unnumbered section headings are generally used for documents that are designed for end users. *Numbered* section headings are typically reserved for hardware installation manuals, technical references, and service manuals.

Numbered headings use a digit for each heading level. The numbers start with the chapter number. The digits are separated by a period.

The section number 4.2.3.1 tells readers that the text is in Chapter 4, the second first-level heading, the third second-level heading, and the first third-level heading. These sections are numbered in this way:

4.2
  4.2.1
  4.2.2
  4.2.3
    4.2.3.1
    4.2.3.2

Lists

Lists are used to break out information from the paragraph format and to structure the information into an easier-to-read format. Lists must include at least two items. Be sure that lists are unmistakably lists. You do not want the reader to confuse a list with steps, which denote actions. Use secondary entries only if you cannot avoid them. Complex entries defeat the easy-to-read list format.

Use *unnumbered (bulleted)* lists when the items are not dependent on the sequence in which you present them. When the items are dependent on sequence, use *numbered* lists with numerals and letters to build the hierarchy.

Use running text when you have up to four one-word items that have equal weight and no order. Incorporate the items in a sentence in which the items are separated with serial commas. You can also put the items in a list if you want to call particular attention to the items.

Use vertical lists in the following situations:

- When you have a list of five or more items
- When you have a list of two or more items in which any of the items consists of two or more words
- When you have a list of two or more items in which any of the items is a link

If a list has more than nine items, try to identify a way to divide the list into two or more lists.

This section covers the following topics:

- "List Introduction Guidelines" on page 63
- "List Introductions and Verbs" on page 65
- "Capitalizing and Punctuating Lists" on page 66
- "Writing Bulleted Lists" on page 67
- "Writing Numbered Lists" on page 68
- "Writing Jump Lists" on page 69

List Introduction Guidelines

This section provides guidelines for constructing list introductions.

- Introduce a list with one of the following constructions:
  - A complete sentence
  - A paragraph
  - A sentence fragment

    A sentence fragment contains a verb. For example:

    The `ab2admin` server management functions include:

    - Stopping the server
    - Starting the server
    - Restarting the server
    - Turning the server log files on or off
    - Rotating the log files
  - A noun phrase

    A noun phrase does not contain a verb. For example:

    Legal requirements checklist:

    - Check that the copyright information is up to date.
    - Ensure that trademarks and service marks are used as adjectives and are properly attributed.
    - Ensure that third-party trademarks are used as adjectives and are properly attributed.
    - Ensure that you properly label confidential information.
- Do not insert a list between the beginning and the end of a sentence.

  Incorrect:

  The calculator can be used for the following operations:

  - Addition
  - Multiplication
  - Subtraction
  - Division

  and has a large, easy-to-read display.

  Correct:

  The calculator has a large, easy-to-read display and can be used for the following operations:

  - Addition
  - Subtraction

- Multiplication
- Division

- When introducing a list, do not construct the list items so that they complete a sentence.

  List items should be grammatically complete.

  Incorrect:

  Use App Builder if you:

  - Are not an expert PlirgSoft programmer
  - Are not familiar with desktop services
  - Want to build your application interface quickly

  Correct:

  Use App Builder in the following situations:

  - If you are not familiar with desktop services
  - If you are not an expert PlirgSoft programmer
  - When you want to build your application interface quickly

- Do not end an introductory phrase with a preposition.

  Incorrect:

  Entities are useful for:

  - Referencing a common string of text
  - Using the same text among different authors
  - Preventing entry errors

  Correct:

  Entities are useful in the following situations:

  - Referencing a common string of text
  - Using the same text among different authors
  - Preventing entry errors

- When introducing a list, use a colon if the introduction clearly anticipates the list, especially if the introduction contains phrasing such as "the following" or "as follows."

  If the introduction is complete in itself, use a period.

  For example, you could use either of these statements to introduce the same list:

  Send a mail message in any of the following three ways:
  The system provides three convenient ways to send a mail message.

- When you use a complete sentence or a phrase with a colon to introduce an unnumbered list, be sure that the syntax of the items in the list is parallel and agrees with the syntax of the introduction.

  Incorrect:

  You can use Mail Tool to perform the following tasks:

  - Compose a new message
  - Replied to a message sent to you
  - Forwarding a message to another person

  Correct:

  You can use Mail Tool to perform the following tasks:

  - Compose a new message
  - Reply to a message that was sent to you
  - Forward a message to another person

List Introductions and Verbs

Avoid ending list introductions with a verb, unless it is an imperative verb. Ending the introduction to a list with a verb usually causes problems for translators and nonnative English speakers.

- Do not end an introductory phrase with the word "to" as part of infinitives in the list.

  Incorrect:

  This section explains how to:

  - Prevent unauthorized users from gaining system access
  - Protect shared servers
  - Create user names and passwords

  Correct:

  This section explains the following tasks:

  - Preventing unauthorized users from gaining system access
  - Protecting shared servers
  - Creating user names and passwords

- Do not end an introductory phrase with a modal verb.

  Modal verbs include "can," "could," "may," "might," "must," "should," "will," and "would."

  Incorrect:

  To do complete print integration, your application must:

  - Provide a print action
  - Use all the environment variables for desktop printing

  Correct:

  Your application must meet the following requirements to do complete print integration:

  - Provide a print action
  - Use all the environment variables for desktop printing

- An introductory phrase can end with an imperative verb.

  In the login screen, type:

  - Your system name
  - Your user name
  - Your password

Capitalizing and Punctuating Lists

For consistency, follow these guidelines when you construct the items in lists:

- Capitalize the first word of each list item in all lists.
- Use punctuation at the end of each item in a list of complete sentences.
- Use no punctuation at the end of each item in a list of sentence fragments.

Avoid mixing complete sentences and sentence fragments in the same list. If you must have a mixed list, add periods at the end of every list item.

To make the list items parallel, a sentence in a mixed list might be preceded by a fragment that describes or introduces the sentence. The fragment ends with a period. The following examples illustrate this recommendation.

Incorrect:

- Fragment
- Fragment
- This item is a complete sentence.
- Fragment

Better:

- Fragment.
- Fragment.
- This item is a complete sentence.
- Fragment.

Best:

- Fragment.
- Fragment.
- Fragment. This item is a complete sentence.
- Fragment.

Writing Bulleted Lists

When the sequence of the list items is not important, use a bulleted (unnumbered) list. Follow these guidelines:

- Make sure that the items in a bulleted list are similar in value.

  Incorrect:

  The workstation that you purchased comes with the following hardware:

  - System unit
  - Monitor
  - Keyboard and mouse
  - Maybe a CD-ROM drive
  - Maybe a modem

  The last two items in the example are not similar to the first three items because they are options, not standard equipment.

  Correct:

  The workstation that you purchased comes with the following hardware:

  - System unit
  - Monitor
  - Keyboard and mouse

  After setting up your workstation, you can add several options, such as a CD-ROM drive or a modem.

- Use a summary word or phrase for each list item, when needed.

  Lists sometimes begin with a summary word or phrase, followed by an explanation. Present the summary word or phrase in bold. The text that follows the summary word or phrase is in the same font as the document's body text. This style is often referred to as a *bold lead-in*. This structure works well when you have explanatory text for each list item.

  After the summary word or phrase for each list item, use either a period or an en dash surrounded by spaces.

  > Select one of the following Find options:
  >
  > - **Change** – Changes the first instance and proceeds to the next instance
  > - **Change All** – Changes all instances
  > - **Skip** – Skips this instance and proceeds to the next instance
  >
  > The workstation that you purchased comes with the following hardware:
  >
  > - **System unit.** This unit houses the main components of the computer.
  > - **Monitor.** This monitor is a 19-inch color monitor.
  > - **Keyboard and mouse.** These input devices are part of the computer package. Different colored covers for the mouse device are included.

  Choose a punctuation strategy for the bold lead-ins that appear in your document or documentation set. Three strategies to consider are as follows:

  - Use only a period or only an en dash throughout your documentation set.
  - Use a period only if the text following the summary word or phrase is extensive. Otherwise, use an en dash.
  - Use a period if the items in the list end with periods. Use an en dash if the items in the list do not end with periods.

Writing Numbered Lists

Use numbered lists when the order of the items is important. However, exercise caution. Many readers are impatient to complete tasks and could mistake numbered lists for procedures. Follow these guidelines:

- Write the text for numbered lists in a style that differs from the style of instructional steps in a procedure.

  Do not assume that a reader will notice any format differences between numbered lists and numbered steps.

- Avoid using verbs in the imperative form.

  Using an imperative verb could lead a reader to believe that the numbered lists are procedures. Use gerunds or participles instead.

  Incorrect:

  To create a file with the `vi` editor, you need to perform the following basic operations:

  1. Start `vi`.
  2. Add text to the file.
  3. Write the file to save its contents.
  4. Quit `vi`.

  To avoid possible misinterpretation, introduce the list clearly and do not use imperative verbs.

  Correct:

  Creating a file with the `vi` editor involves the following basic operations:

  1. Starting `vi`
  2. Adding text to the file
  3. Writing the file to save its contents
  4. Quitting `vi`

Writing Jump Lists

A *jump list* is a bulleted list of cross-references that serves as a kind of table of contents for a portion of a book, usually a chapter or a long section. Jump lists are useful navigational tools for a reader. These lists also provide you with a good test of the integrity of your document structure. If your document is online and hypertext is available, include links to the selected sections.

Note – If your document is going to be presented in a specific online display, check whether jump lists are generated automatically as part of the navigational system.

When constructing jump lists, follow these guidelines:

- Try to keep the text of a jump list short enough to fit on one page.
- Try to use jump lists consistently in your document.

  If you use a jump list at the beginning of one chapter, try to begin every chapter with a jump list. If you use a jump list at the beginning of one long section, try to begin each long section with a jump list.
- List headings for topics that might be of particular interest to a reader.

  You do not need to include only first-level headings, but be consistent throughout a book.

- Whenever possible, place introductory text before the jump list at the start of a chapter. If you are using jump lists for online navigation, readers might not notice text placed after the list.

  After the jump list, you might include a cross-reference to another book or chapter.

Tables

Tables are an ideal format for presenting statistical information or facts that you can structure uniformly. Tables must include at least two rows. Information that is conceptual or explanatory is best written in a narrative paragraph.

Writing Text for Tables

Tables typically include a number, caption, column headings, and table text. Use spaces and rules (vertical and horizontal lines) to format the table text, when necessary.

Table Introductions

If you need to introduce the context of a table to your readers, follow these guidelines:

- Use a complete sentence ending with a period, rather than a colon.
- Refer to the table's position on the page in the document flow, for example, "The following table describes the compatible applications."

  In subsequent references to the same table, use "preceding" as appropriate.
- Refer to the table number only if the table does not immediately follow or precede the paragraph that contains the reference.

  In online documents, references to table numbers typically become active links. Readers expect links to take them to another section of the document or to another page. Therefore, a link that takes them nowhere or moves only a slight increment down the page is potentially frustrating or disorienting.

- Do not insert a table between the beginning and the end of a sentence.

  Incorrect:

  Add the following components from your installation CD

Component	Description
Tutorial	Includes step-by-step introductory procedures to help you get started
Templates	Common business document structures that you can customize
Spelling checker	An interactive spelling checker

  and their related updates.

  Correct:

  Add the following components and their related updates from your installation CD.

Component	Description
Tutorial	Includes step-by-step introductory procedures to help you get started
Templates	Common business document structures that you can customize
Spelling checker	An interactive spelling checker

Table Captions

When you construct table captions, do the following:

- Use formal tables with numbers and captions, except when the context of the tables is so clear that the numbers and captions are redundant.
- Capitalize table captions.

  Match the table caption style to the figure captions and section headings in your document.
- Decide how you can indicate when table text runs onto more than one page in the printed document.

  "Continued" usually appears in a caption on the second and subsequent pages. For longer tables, you might indicate how many pages the table runs and the current page in that sequence, for example, "Sheet 3 of 8." Your authoring environment might insert this wording for you.

Table Column Headings

Table column headings concisely summarize information in a column. To create table column headings, follow these guidelines:

- Avoid starting a column heading with an article.

  For example, write "Alternative Backup Schedule" rather than "An Alternative Backup Schedule."
- Do not use end punctuation, except a question mark.
- Do not use sentence fragments with verbs in which the verb is in the column heading and the object of that verb is in the body of the table.

  For example, "For More Information, Go To" contains a verb and the object of that verb is the cross-reference to the additional information. In this case, a better heading would be "Further Information."

  Incorrect:

With This Method...	You Can...
PlirgSoft Quick Start	Install all the software in your product box (the PlirgSoft software and co-packaged software) at once from a single, browser-based tool.

  Correct:

Installation Method	Features Supported
PlirgSoft Quick Start	Install all the software in your product box (the PlirgSoft software and co-packaged software) at once from a single, browser-based tool.

- Capitalize column headings as you do section headings.

  Capitalize the first letter of the first word and the first letter of all other words *except* conjunctions, articles, prepositions of fewer than four letters, and the "to" in infinitives.

Table Text

The table text is the main body of information, formatted into rows and columns. To construct table text, follow these guidelines:

- Use parallel construction, capitalization, and punctuation.
  - For better readability, use an initial capital for only the first word in a table cell unless a reason exists to capitalize other words in the text.
  - Use a period at the end of each entry in a column when one or more entries are complete sentences. Try to use either all phrases that do not require periods or all sentences that do require periods.
- Write table text as concisely as possible.

- You may use abbreviations or symbols in tables when you need to conserve space.

  Some common space savers include "no." for "number" and "%" for percentages.
- Avoid bold in table text, but use the typographic conventions established for the document.
- For footnotes in a table, use numerals when possible.

  When numeric footnotes might be confusing due to numerals in the table text, use these symbols instead in the following order:
  - Asterisk (*)
  - Section mark (§)
  - Double asterisk (**)
  - Double section mark (§§)

Determining the Type of Table to Use

Tables present information in concise categories. The way that you design a table depends on the information that you need to present.

Note – The HTML version of this style guide shows tables with horizontal and vertical lines, regardless of how the table is designed. The printed version accurately reflects how the different tables should appear.

The following example shows a standard table with columns and rows separated by spaces.

EXAMPLE 2–1 Standard Table

Specifier	Value of the Variable	Data Type for the Variable
ACCESS	'DIRECT SEQUENTIAL'	CHARACTER
BLANK	'NULL' 'ZERO'	CHARACTER
IOSTAT	Error number	INTEGER
OPENED	.TRUE. .FALSE.	LOGICAL

The following table uses horizontal lines to group information into rows.

EXAMPLE 2–2 Horizontal Lines Table

Name and Address	Corporate Office	Sales	Service
ABC Corp. 624 Main Street Chelmsford, MA 01824	+1-617-555-9731	+1-617-555-1632	+1-617-555-4932
DEF Corp. 90 Columbia Avenue Los Angeles, CA 94043	+1-213-555-8413	+1-415-555-5940	+1-415-555-3662
GHI Corp. Colorado Springs, CO 80920	+1-719-555-8842	+1-719-555-9013	+1-719-555-4701

The following table uses horizontal and vertical lines to separate information.

EXAMPLE 2–3 Horizontal and Vertical Lines Table

Command	Syntax	Options
`at`	`at` *time* `at` [*options*] *job-IDs*	`-l` Lists current job. `-r` Removes specified job. *time* Specifies the time when commands will run.
`chown`	`chown` *owner filename*	`-h` Changes ownership of symbolic link.
`find`	`find` *filename*	`-print` Prints names of files found. `-name` *filename* Finds file with cited name.

The following table uses side and top headings to create a grid. Both the top and side headings are bold. A vertical line separates the side headings from the table text.

EXAMPLE 2–4 Side and Top Headings Table

Permission	User	Group	Others
Read	4	4	4
Write	2	0	0
Execute	1	1	1
Total	7	5	5

The following table uses a two-column format. You can use this format as an informal table without a number or a caption. Two-column tables are often used for command options and their descriptions. In some authoring environments, this format might be a list rather than a table.

EXAMPLE 2–5 Two-Column Table

`-a`	Lists hidden files.
`-l`	Includes date and size information in the file list.
`-r`	Lists the files in reverse alphabetical order.

Code Examples

Code examples are portions of computer programs that you include in a document to help explain a topic. Code examples can include the dialogue between a user's input and the computer's responses. Examples can also include only the code that a person types into the computer.

Because programming code is precise, you must reproduce the exact code, even if the code contains language errors in spelling, grammar, or punctuation. However, you can correct the *comments* in code, even if doing so means that the example will be inconsistent with the actual source code. If you have the opportunity, bring any errors to the attention of the person or group supplying the code so that the errors can be corrected in the source.

Use monospace font for code examples.

To include a code example in a List of Code Examples in the front matter, add a caption. Capitalize and punctuate the caption as you would any other caption, title, or heading.

Note – If you are presenting lengthy programs, put them in an appendix and cross-reference them unless the entire book consists mainly of long code examples.

For legal reasons, be careful when providing examples of code. See "Protecting Confidential Information in Examples" on page 200.

Error Messages

When documenting error messages, follow these guidelines:

- Reproduce the exact message text, even if the text contains language errors in spelling, grammar, or punctuation.

  If you have the opportunity, bring the error to the attention of the person or group supplying the error message text so that the error can be corrected in the source.

  Use monospace font for the error message text.
- Try to provide troubleshooting information when discussing error messages, such as why the message has appeared and what a user can do to correct the problem.

  Such explanations are not required when error messages are presented in running text.
- If you are documenting numerous error messages, consider compiling them into an appendix or even into a separate book.

The following examples show some ways that you can format error messages in text and in a table, depending on your needs and your authoring environment.

EXAMPLE 2–6 Formatting Error Messages as Text

`Current working directory,` *directory*`, not a valid installation area`

> **Cause:** You are running the installation program from a directory other than where the `install` package was installed.
>
> **Solution:** You must change to the directory in which the `install` package was installed and run the `install` package again.

EXAMPLE 2–7 Presenting Error Messages in a Table

Message		Cause	Solution
`6000`	`Signal to Noise ratio too low on` *name*, SNR = *number* db, `Min` SNR = *number* db (*text*)	■ Loopback cable is missing or faulty. ■ Audio hardware problem (usually consistent failures). ■ System software problem (usually intermittent failures).	Contact system administrator for assistance.
`8000`	`Must be superuser to execute. The user does not have superuser privileges.`	User does not have superuser privileges.	Contact system administrator for superuser privileges.
`8012`	`Invalid audio device` *device-name* `for Crystal test`	Crystal test is not supported on system audio.	Provide required equipment.

Cross-References

Cross-references identify additional information about a specific topic that is available in the document or elsewhere. In online documents, cross-references become links. To be useful to a reader, cross-references must be specific and accurate. Include all details that can help a reader find the information easily.

For information about how to write links for online documents, see Chapter 6, "Constructing Links."

Several formats are acceptable for cross-references. These formats often depend on the location of the cited information, the length of the reference, and your authoring environment. The cited information can be in your document, another document produced by your company, or a third-party document.

Do *not* use a cross-reference in the following situations:

- When the information is vital for a reader to understand the discussion

  Instead, provide the vital information. If the information is extensive, you can summarize it and also include a cross-reference to the source.
- When the additional information is brief and you can just as easily repeat it
- When you cite safety information that describes how to protect a person, hardware, or software

  For safety information, use Caution text. For more information, see "Writing Cautions" on page 83.

Formatting Cross-References

The punctuation, capitalization, and fonts that are used for cross-references might be determined by your authoring environment. Otherwise, follow these guidelines:

- Use italic for the title of a book, journal, multivolume work, or magazine.
- Use quotation marks or commas around a chapter title or section heading.

  See "Setting Administrative Options" on page 97.

  See Chapter 4, "System Board and Component Replacement," in the *Plirg 5769 System Service Manual.*
- Capitalize "chapter," "appendix," "part," "section," "table," "figure," "example," or "step" when these terms are followed by a number or letter.

  See Chapter 6.
  This process is shown in Section 3.5, "Null Modem Cabling."
- Do not capitalize the word "page" when used in a page number reference.

  Refer to page 42 for further information.

Writing Cross-References

Cross-references break the flow of your discussion. Therefore, write cross-references so that a reader easily recognizes when you have given a reference.

- Make sure that you introduce cross-references with clear phrases.

  Incorrect: To reboot, see Chapter 4.
  Correct: For instructions on how to reboot, see Chapter 4.

- If cross-references are brief, include them within sentences.

  Use the `diff` *filename1 filename2* command to compare two files (see "Working With Text Files" on page 86).

  If the cross-references require lengthy text references, put the cross-references in separate sentences.

  Use the `diff` *filename1 filename2* command to compare two files. See Chapter 2, "Working With Text Files," in *UNIX Simplified.*

- Do not provide the title of your document as part of the cross-reference.
- Do not provide a cross-reference to a figure, table, or example that appears adjacent to the paragraph text.

  Instead, provide the relative location, for example, "The following figure shows the File menu."

  In online documents, cross-references to figures, tables, and examples typically become active links. Readers expect links to take them to another section of the document or to another page. Therefore, a link that takes them nowhere or moves only a slight increment down the page is potentially frustrating or disorienting.

- For a cross-reference to a third-party book, include the title and author, with the publisher and year in parentheses.

  One of the reference books used in preparation of this document was *The Deluxe Transitive Vampire* by Karen Elizabeth Gordon (Pantheon Books, 1993).

  The title is sufficient for subsequent references in the same chapter.

Endnotes, Footnotes, and Bibliographies

Sometimes, you need to provide complete cross-references to other sources. You can include these references in *endnotes* at the end of a chapter, in *footnotes* at the bottom of a page, or in a *bibliography* at the end of a document. You also can use endnotes and footnotes as a place to add comments about a discussion.

Writing Endnotes and Footnotes

Endnotes and footnotes provide complete information about the source, including author, title, place of publication, and page number in the source where readers can find the information. A footnote provides the reference at the bottom of the page, while an endnote groups all references cited in a chapter at the end of the chapter.

When writing endnotes and footnotes, follow these guidelines:

- If an online presentation or authoring environment restricts the use of footnotes to tables only, do one of the following:
  - Incorporate the footnote information into the text.
  - Put the footnote information in an endnote.
- Use consecutive superscript numerals in the main text to indicate the endnotes or footnotes.

  Place the numeral at the end of the sentence, phrase, or quotation for which you are providing a reference. Do not include a period after the superscript numeral. The numeral appears after punctuation marks, except for a dash.
- Do not use lettered subnotes, such as 4a and 4b.
- Do not place endnote or footnote numerals in chapter titles or section headings, or in figure, table, or example captions.
- Keep the text of the footnote on the same page as its superscript numeral unless your authoring tool determines the placement for you.
- Introduce the text of the note with the corresponding numeral, followed by a period if the numeral is not a superscript.

  Use consistent alignment of the numerals and text. Use single-spacing for the text of each note.
- When repeatedly referring to the same source document, you can abbreviate the information.

  Include only the author's last name, the book title, and the page number.

The following examples show the content and style of endnotes and footnotes.

1. Skillin, Marjorie E., Robert M. Gay, and other authorities. *Words into Type*, 3d ed. Englewood Cliffs, N.J.: Prentice-Hall, 1974, p. 150.
2. Burnett, Rebecca E., *Technical Communication*, 6th ed. Belmont, Calif.: Wadsworth Publishing Company, 2004, p. 36.
3. *The Chicago Manual of Style,* 15th ed. Chicago: University of Chicago Press, 2003, p. 701.
4. Skillin, *Words into Type*, p. 16.

Writing Bibliographies

A bibliography lists all sources to which you refer in your document. Bibliographies appear at the end of the document, after the appendixes and glossary and before the index.

The format for bibliographies varies, depending on the number and type of resources that are cited. These resources might include books, journals, articles, and so forth. *The Chicago Manual of Style* provides a lengthy discussion on text and formats used in bibliographies. That discussion is summarized in the next few sections, followed by guidelines describing how to cite and format electronic source material.

Citing Books

When citing a book, provide the following information:

- Name of the person or institution credited as author, which could be single or multiple authors, editors, or institutions
- Full title of the source, including any subtitle
- Title of the series, and its volume and number if the book is part of a series
- Volume number if the book is part of a multivolume set
- Edition of the book if not the original edition
- City where the book was published
- Publisher's name
- Date of publication

Citing Articles

For an article in a periodical, provide the following information:

- Name of the author
- Title of the article
- Name of the periodical
- Volume or issue number of the periodical
- Date of the periodical
- Page numbers where the article is found in the periodical

Formatting Bibliographies

Two examples of ways to format text for bibliographies follow. The first example wraps each line flush left, while the second example provides a two-column format.

Gove, Darryl. *Solaris Application Programming*. Upper Saddle River, NJ: Prentice Hall, 2008.

Furht, Borko. *Encyclopedia of Multimedia*. New York, NY: Springer, 2006.

Citing Electronic Sources

When citing electronic source material, provide the following information:

- Name of the author, plus the author's email address if available
- Title of the work or the title line of the message
- Title of the list or site, as appropriate
- Volume or issue number of a digest if applicable
- Date of the message or the date on which the source was accessed

Formatting Electronic Sources

Suggested ways to format various types of electronic sources are shown here:

- Database online

  Joe User, "Citing Electronic Material," in REFSTUFF [database online] (Silicon, Calif.: REFSTUFF, 1986) [updated 9 January 2009; cited 31 February 2009], identifier no. Q000307, [52 lines].

- Email message

  Joe User [`juser@macland.org`], "Citing Electronic Material," private email message to Sys Admin [`sadmin@funix.com`], February 21, 2009.

- FTP site

  Joe User [`juser@macland.org`], "Citing Electronic Material," [`ftp://ftp.macland.org/pub/local/reference/cites.txt`], February 2009.

- Internet article based on a print source

  VandenBos, G., Knapp, S. & Doe, J. (2008). "Selection of Resources by Psychology Undergraduates" [electronic version]. *Journal of Bibliographic Research, 5*, 117–123.

- Article in an Internet-only journal

  Fredrickson, B.L. (2008, November 7). "Cultivating Positive Emotions." *Prevention & Treatment 3*, Article 0001a. Retrieved March 20, 2002 from `http://www.journals.apa.org/prevention/pre110100001a.html`

- Listserv message

  Joe User [`juser@macland.org`], "REPLY: Citing Electronic Material," in GRAMR (Digest vol. 6, no. 8) [electronic bulletin board] (Dubuque, Iowa, 2005) [cited February 21, 2002]; available from [`gramr@miskatonic.edu`]; INTERNET.

- Material from an unknown author

  GVU's 8th WWW user survey. (n.d.) Retrieved April 8, 2009, from `http://www.cc.gatech.edu/gvu/usersurveys/survey2007-09`

- World Wide Web

  Joe User, "Citing Electronic Material," [`http://www.macland.org/cites.html`], accessed February 21, 2009.

Notes, Cautions, and Tips

Notes, Cautions, and Tips provide important information that diverges from the topic under discussion.

A *Note* usually provides information that is related to the text. A Note might contain an explanation, a comment, a reinforcement of the text, or a short expansion of the concepts in the text. A Note might also contain a statement that is intended to catch the reader's attention.

A *Caution* is mandatory text that you *must* provide to protect the user of equipment from personal injury or to protect hardware or software from damage. Precede the text of a Caution with the appropriate graphical symbol specified by the International Organization for Standardization (ISO).

A *Tip* describes practical but nonessential information that does not otherwise fit into the flow of the text.

Try to limit the number of Notes and Cautions to approximately two on a page. The inclusion of many Notes might indicate an organizational problem in the text.

Writing Notes

Few constraints exist on the text or format you can use in a Note. Consistency in both writing style and format is important so that a reader learns to recognize a Note that is interjected into the text.

When writing Notes, follow these guidelines:

- Use a Note to break out related, reinforcing, or other "special" information.
- Keep your Note short and relevant.

- Never use a Note to cite safety information.
- If your text requires many Notes, consider reorganizing the text to reduce the number of Notes that are required.

An example of a Note and its format follows.

Note – Keep the text short and relevant.

Writing Cautions

Unlike a Note, a Caution is not optional. You must provide a Caution in the following situations:

- When you describe a situation that has the potential to cause injury to a person, or when there is a risk of irreversible destruction to data or the operating system
- When you describe anything that has the potential to cause damage to equipment, data, or software

When writing or formatting Cautions, follow these guidelines:

- Be direct when writing a Caution.

  First describe the potential hazard to data, equipment, or personnel. Then, describe the actions that are required to avoid the hazard.
- Insert the Caution before the information that might cause the potential hazard.
- Use the "lightning bolt" symbol when there is danger of physical harm to a person or damage to equipment due to an electrical hazard.
- Use the "heat" symbol when there is risk of personal injury from a heat source.
- Use the "exclamation point" symbol when there is risk of personal injury from a nonelectrical hazard or risk of irreversible damage to data, software, or the operating system.

Examples of Cautions and their graphical symbols follow.

Caution – Lithium batteries are not customer-replaceable parts. Do not disassemble them or attempt to recharge them.

Caution – Software hazard is present. Before copying data from a disk from an outside source, check the disk for viruses. Otherwise, you could contaminate the data on your hard disk.

Caution – Irreversible destruction can occur to data or the operating system. Follow the instructions carefully.

Caution – Electrical hazard is present. Leaving the side panel off the computer exposes you to dangerous voltage and risk of electrical shock. Do not leave the side panel off while you are operating the computer.

Caution – Hot surface. The surface of the CPU chip can be hot and could cause personal injury if touched. Do not touch this component.

Writing Tips

A Tip describes practical but nonessential information that does not otherwise fit into the flow of the text. Examples include a keyboard shortcut or an alternative way to perform a step in a procedure.

When writing Tips, follow these guidelines:

- Keep the text short and relevant.
- Do not introduce new topics or elaborate on conceptual material in a Tip.

An example of a Tip and its format follows.

Tip – You can also minimize a window by pressing Alt-F9.

Key Name Conventions

A *key name* indicates which key on a keyboard to press to obtain a desired action. Follow these guidelines when referring to the names of keys, unless the product you are documenting specifies other requirements:

- Capitalize the name of a key that is a letter.

  Press the A key
  Press Stop-A

However, if you type a letter as part of a procedure, use lowercase monospace font.

1. Type `n` for no.

- Include the Shift key when instructing users how to obtain a capital letter, symbol, or punctuation mark that appears on the upper half of a keycap.

  Press Control-Q Shift-P to include an en dash in a FrameMaker document.
  Press Shift-8 to type an asterisk.

- Use initial capitalization for each term in a key name.

  Press Return
  Press Caps Lock

- Use the short form for key names when three keystrokes are required to perform one action, as shown in Table 2–1.

  Ctrl-Alt-Del

  Notice that the accepted way to refer to some keys is in abbreviated form.

  Num Lock (the Numerical Lock key)
  Sys Req (the System Request key)

Use the previous guidelines and the format of key names in the following table if the product assigns different names to keys. For example, if F7 is renamed Pick Copy, this key name would appear as two words with initial capitals.

Referring to Keys

The following table shows how to refer to common names of keys. Even if Ctrl, CONTROL, or CTRL appears on your keyboard, for example, refer to that key as the Control key. Use the short form when noting three simultaneous or consecutive keystrokes or when using the key names in tables.

TABLE 2–1 Key Name Conventions

Standard Name	Short Form	Standard Name	Short Form
Again		left arrow (no caps)[1]	
Alt		Meta (for <>)	
Alt Graph		Num Lock	
Backspace		Open	
Backtab		Page Down	PgDn (no space)

[1] If an arrow key is used in a keystroke combination, use initial capitals, for example, Shift-Down Arrow.

TABLE 2–1 Key Name Conventions *(Continued)*

Standard Name	Short Form	Standard Name	Short Form
Break		Page Up	PgUp (no space)
Caps Lock		Paste	
Compose		Pause	
Control	Ctrl	Print Screen	PrtSc (no space)
Copy		Props	
Cut		Return	
Delete	Del	right arrow (no caps)[1]	
down arrow (no caps)[1]		Scroll Lock	
End		Shift	
Enter[2]		spacebar (no caps)	
Escape	Esc	Stop	
F1 to F10 or F12		Sys Req	
Find		Tab	
Front		tilde (no caps) (for ~)	
Help		Undo	
Home		up arrow (no caps)[1]	
Insert	Ins		

[1] If an arrow key is used in a keystroke combination, use initial capitals, for example, Shift-Down Arrow.

[2] When writing about personal computers, use Enter, not Return.

Documenting Multiple Keystrokes

Consider using the following conventions to describe multiple keystrokes unless you are documenting Microsoft products:

- Use a hyphen to join *simultaneous keystrokes*.

  Press the first key while you press subsequent keys.

  Control-A
  Ctrl-Shift-Q
  Meta-A

- Use a plus sign to join *consecutive keystrokes*.

  Press the first key, release it, then press subsequent keys.

  Ctrl+A+N
  F4+Q

If you are documenting Microsoft products, follow these conventions:

- Use a plus sign to join simultaneous keystrokes with no spaces around the plus sign, as in "press Control+F."
- Use a comma and a space to separate each key name in consecutive keystrokes, as in "press Alt, F, N."

Follow alternative key name conventions as appropriate for your audience.

Symbol Name Conventions

The following table shows how to refer to symbol names that typically appear in technical documentation. Common symbols with only one generally accepted name are not included. For example, the at sign (@) and the comma (,) are not included in this table for this reason.

TABLE 2–2 Symbol Name Conventions

Symbol	Standard Name	Notes
&	Ampersand	Not "and."
*	Asterisk	Not "star."
\	Backslash	
{}	Braces	Not "curly brackets" or "curly braces."
[[]]	Double square brackets	Not "double brackets."
=	Equal sign	Not "equals sign."
!	Exclamation point	Not "exclamation mark" or "bang."
≥	Greater than or equal to sign	

TABLE 2–2 Symbol Name Conventions *(Continued)*

<table>
<tr><th>Symbol</th><th>Standard Name</th><th>Notes</th></tr>
<tr><td>></td><td>Greater than sign</td><td></td></tr>
<tr><td>≤</td><td>Less than or equal to sign</td><td></td></tr>
<tr><td><</td><td>Less than sign</td><td></td></tr>
<tr><td>–</td><td>Minus sign</td><td></td></tr>
<tr><td>×</td><td>Multiplication sign</td><td></td></tr>
<tr><td>≠</td><td>Not equal to sign</td><td></td></tr>
<tr><td>¶</td><td>Paragraph mark</td><td>Not “carriage return.”</td></tr>
<tr><td>()</td><td>Parentheses (pl), opening or closing parenthesis (s)</td><td></td></tr>
<tr><td>%</td><td>Percent sign</td><td></td></tr>
<tr><td>.</td><td>Period or dot</td><td>When used as punctuation, use “period.” When used as a separator such as in URLs and file names, use “dot.”</td></tr>
<tr><td>±</td><td>Plus or minus sign</td><td></td></tr>
<tr><td>+</td><td>Plus sign</td><td></td></tr>
<tr><td>#</td><td>Pound sign, comment mark, or superuser prompt</td><td>Which term to use depends on the context in which this symbol is used.</td></tr>
<tr><td>“”</td><td>Quotation marks</td><td>Not “quotes” or “quote marks.”</td></tr>
<tr><td>§</td><td>Section mark</td><td></td></tr>
<tr><td>''</td><td>Single curly quotation marks</td><td>Not “quotes” or “quote marks.”</td></tr>
<tr><td>' '</td><td>Single straight quotation marks</td><td>Not “quotes” or “quote marks.”</td></tr>
<tr><td>/</td><td>Slash</td><td>When necessary to distinguish from a backslash, use “forward slash.” Otherwise, use “slash.”</td></tr>
<tr><td>[]</td><td>Square brackets</td><td>Not “brackets.” For information on this symbol, see “Square Brackets” on page 56.</td></tr>
<tr><td>""</td><td>Straight quotation marks</td><td>Not “quotes” or “quote marks.”</td></tr>
<tr><td>_</td><td>Underscore</td><td>Not “underbar” or “underline.”</td></tr>
<tr><td>|</td><td>Vertical bar or pipe</td><td>Use “vertical bar,” except when the symbol is used to pipe output from one command to another command. Then, use “pipe.”</td></tr>
</table>

◆ ◆ ◆ CHAPTER 3

Writing Style

If *content* is *what* you communicate, then *style* is *how* you communicate. Writing style is determined by all the decisions that you make while creating a document, such as the type and tone of information you present, choice of words, language and format consistency, use of technical terms, and so forth. In the literary world, style is judged in part on artistic grounds, which might be highly subjective. In the field of technical documentation, however, experience and practice have provided objective criteria for evaluating style.

This chapter presents some guidelines for writing effectively. It discusses the following topics:

Why Is Style Important?

Good style is synonymous with effective communication. Documents that communicate effectively reduce costs and increase customer satisfaction. Style that responds to the requirements of readers results in fewer revisions, fewer calls to customer support, reduced training needs, and easier translation. Customer satisfaction increases when accurate and functional documentation enables customers to use a product quickly and efficiently.

Stylistic Principles

Keep in mind a few stylistic considerations when writing computer documentation: simplicity, accuracy, and consistency. Two principles underlie stylistic considerations:

- **Time is a valuable commodity.** Readers of computer documentation are generally in a hurry. Readers turn to documentation to find answers to problems and are impatient to get on with the task at hand. Write in a style that aids the customer's speedy understanding of the product.
- **Readers are worldwide.** International markets are a significant source of revenue. Documentation is being translated more frequently than ever before.

For more information about style and internationalization, see Chapter 8, "Writing for an International Audience."

Write Simply, Directly, and Accurately

People most often read technical manuals to find answers to problems that they are having with software or hardware. They need their questions answered concisely and accurately. Concise writing means readers do not have to contend with unnecessary technical jargon. Write simple and direct sentences. Use short, familiar words, but respect the reader's level of technical knowledge and competency. Simple, direct, and accurate writing makes a document more usable and easier to translate.

Be Consistent

Readers project some significance onto every change in tone, language, or typographic convention. A consistent style enables readers to internalize the language and text conventions of a document. As a result, understanding occurs more easily and significant points stand out more clearly. Consistency is one of the most valuable aspects of good style.

Some Basic Elements of Style

At every level of writing, you must make stylistic decisions, from word choice to paragraph structure. The following sections discuss a few aspects of these style decisions.

Avoid Jargon

Writers frequently incorporate jargon associated with the subject matter into their documentation. *Jargon* is the specialized language of a profession, which is often meaningless to outsiders. Jargon can be difficult for the "uninitiated" to understand. In addition, jargon can be very difficult to translate.

Awkward: Disambiguate the code.
Better: Clarify the code.

When you have to use computer terms, introduce them in italic, explain them, include them in a glossary, and use them consistently.

Use Active Voice and Passive Voice Appropriately

Always try to write in the active voice, but do not fear the correct, thoughtful use of the passive voice. Writing entirely in the active voice is nearly impossible to achieve, so know when to use the passive voice.

Because writers in the computer industry often insist on using the active voice, the writer of the following example introduces a message with this sentence:

> As soon as the application completes, the following message displays.

Rewrite: A message does not display. Rather, a message appears or is displayed by the system. The passive voice can indicate that the subject is the receiver of the action rather than the performer.

> As soon as the application is finished, the following message is displayed.
>
> Or
>
> When the application finishes, the following message appears.

Make your writing active by concentrating on the activity of your subject. Use the passive voice when it is unavoidable because the performer of the action is either unimportant to the reader or unidentifiable.

Use Present Tense and Future Tense Appropriately

Readers use technical documents to perform tasks or gather information. For readers, these activities take place in the present. Therefore, the present tense is appropriate in most cases. Only use the future tense when necessary.

Incorrect:

If you attempt to copy a directory without using this option, you will see an error message.

Correct:

If you attempt to copy a directory without using this option, you see an error message.

Use Sentence Structures That Enhance Understanding

Convoluted sentences or sentences densely packed with information cause confusion, slow the reading process, and are difficult to translate. Any sentence that attempts to convey too much information is too long, regardless of its word count. Use punctuation, rhythm, and clarity of meaning to regulate sentence length and to attain a style that is easy to understand.

Write as if you were talking to a person, rather than formulating a law or theorem. Consider the following example:

> To scroll directly to a relative location in the document, move the pointer into the bar at the point that represents the relative location of the text in the document.

Comments: A person would never speak that sentence. The sentence is grammatically correct, but it is stiff and formal. Ask yourself, "Does this sound like me responding to a question?"

To get a response such as the previous example, a person would have had to ask, "How can I scroll directly to the relative location of text in my document?" More likely the person would ask, "If what I want to read is way up or way down in the document, how can I get to it without scrolling through every line?"

Rewrite: Respond naturally to the reader's question. Remember that the reader is a person who wants to do something.

> If the text that you want to read appears elsewhere in the document, guess where it is. Then, move the pointer to that spot in the scroll bar, and click.

Avoid Complex, Conjoined Sentences

A long, complicated sentence that contains several concepts is difficult to translate and to understand. Try to keep sentences to one idea. Rewrite the sentence or break the sentence into several shorter sentences.

> **Incorrect:**
>
> The descriptions in this chapter follow the flow of data through an organization, starting with the back-end data repositories and working through them to the user-access layer provided by the web server, making the assumption that these components are connected by a reliable, available, and scalable network infrastructure.
>
> **Correct:**
>
> The descriptions in this chapter follow the flow of data through an organization. The flow begins with the back-end data repositories. Data then works through the repositories to the user-access layer provided by the web server. In these descriptions, the components are connected by a reliable, available, and scalable network infrastructure.

Also, a sentence that contains more than two uses of "and" or "or" can be difficult for readers to understand. Readers have difficulty with such sentences when multiple conjunctions join more than one main idea.

Incorrect:

From the Addresses tab, you can add or delete networks, and add or delete IP addresses individually or in blocks.

Correct:

From the Addresses tab, you can perform the following operations:

- Adding or deleting networks
- Adding or deleting IP addresses individually or in blocks

Separate Independent Clauses Appropriately

Readers can parse simple sentences more easily than compound sentences. Therefore, avoid combining independent clauses with "and." Instead, write two separate sentences.

Incorrect:

The Motif program uses Motif Version 2.1, and the old shared library uses Motif Version 1.2.

Correct:

The Motif program uses Motif Version 2.1. The old shared library uses Motif Version 1.2.

Limit Subordinate Clauses

Readers can have difficulty parsing sentences that contain a number of subordinate clauses. Limit subordinate clauses, such as "She said that Kathy said that she updated the file."

Use Positive Constructions

Negative constructions can cause confusion. Use positive constructions to state advice or instructions.

Incorrect:

You cannot reconnect to the server without restarting your computer.

Correct:

Restart your computer to reconnect to the server.

Use Parallel Structure

When you use "and" or "or" to link phrases, the reader expects parallelism on both sides of the conjunction. Be sure that your linked phrases are of the same type, for example, noun phrase or verb phrase.

Incorrect:

You can use Mail Tool for composing and to send messages.

Correct:

You can use Mail Tool to compose messages and to send them.

Avoid the Subjunctive

To avoid confusion by readers unfamiliar with the subjunctive mood, use the indicative mood instead where possible.

Subjunctive:

The procedure requires that the packages be first installed.

Indicative:

Before performing this procedure, ensure that the packages are installed.

Differentiate Between Restrictive Elements and Nonrestrictive Elements

A restrictive or nonrestrictive element can be either a clause or a phrase. Make sure that you distinguish between restrictive and nonrestrictive elements. Consider the differences in meaning for the following two sentences:

Check the LED that is on the front panel. (restrictive clause)
Check the LED, which is on the front panel. (nonrestrictive clause)

In the first sentence, the reader is told to check the LED specifically on the front panel, not the one on the side panel or back panel.

In the first part of the second sentence, the reader is told merely to check the LED. The second part of the sentence also states that the LED happens to be on the front panel. This clause implies that no other LED exists anywhere else. The minor difference in meaning could confuse translators or nonnative speakers of English.

Make sure that you include the word "that" when introducing a restrictive clause.

Incorrect:

This chapter provides the information you need to install the software.

Correct:

This chapter provides the information that you need to install the software.

Divide nonrestrictive clauses that are associated with a relative pronoun into separate sentences. This separation can help the translator to understand the meaning.

Unclear:

This topic describes how to write makefiles that take full advantage of CodeManager and ParallelMake, the make utility that is included with the PlirgWare release.

Clear:

This topic describes how to write makefiles that take full advantage of CodeManager and ParallelMake. ParallelMake is the make utility that is included with the PlirgWare release.

Construct Scannable Paragraphs, Headings, and Lists

Readers expect text to be succinct and directly relevant. They are also goal-oriented and scan for information rather than read long blocks of text thoroughly. Make documents scannable to help readers locate information quickly.

To construct scannable paragraphs, headings, and lists, do the following:

- Write clearly and simply.
- Keep paragraphs short.
- Condense text.

  Write concisely and eliminate unnecessary material. However, be cautious about condensing text too much. See "Be Careful When Condensing Text" on page 99.
- Replace text with tables, charts, and figures, when possible.
- Write meaningful headings and subheadings.
- Use bulleted lists and jump lists.

Write Clearly and Simply

To present information clearly and simply, follow these guidelines:

- Use simple declarative and imperative sentence structures.
- Use active voice, present tense, and concrete, meaningful words.

- Use terms consistently.

  When you use terms consistently, readers do not have to reread different sections to grasp the meaning of the material.

EXAMPLE 3–1 Writing Clearly and Simply

Incorrect:

It is recommended that virtual memory limits be set high so that these limits can never be reached under normal operating circumstances. However, the virtual memory limit can also be used to place a limit over the entire database server in order to stop a failing database with a memory leak from spilling over to other databases or workloads on the system.

Correct:

Set virtual limits high so that they are never reached under normal circumstances. However, you can use the virtual memory limit to place a limit over the entire database server. This strategy prevents a database with a memory leak from affecting other databases or workloads on the system.

Keep Paragraphs Short

Short paragraphs offer visual breaks and are easier to scan and read. To keep paragraphs short, follow these guidelines:

- Limit each paragraph to one idea.
- Describe the main topic of the paragraph in the first sentence.
- Emphasize main points.

  Set off each main point in its own sentence or short paragraph.
- Keep paragraphs from three to five sentences long.
- Limit paragraphs to 75 words or fewer.
- Avoid paragraphs that are so dense with information that the reader must struggle to understand the information.

EXAMPLE 3–2 Keeping Paragraphs Short

In the following example, a long paragraph is replaced by a list.

Incorrect:

There are two methods for setting device attributes. The first method is to call the `xil_device_create` and `xil_device_set_value` functions before creating the device image with the `xil_create_from_device` function. The second method is to call `xil_device_set_value` after calling `xil_create_from_device`. Certain attributes require that they be set with the first method (such as `DEVICE_NAME`) and are documented as such in the following sections.

EXAMPLE 3–2 Keeping Paragraphs Short *(Continued)*

Correct:

You can use one of these methods to set device attributes:

- Call the `xil_device_create` and `xil_device_set_value` functions before creating the device image with `xil_create_from_device`.
- Call `xil_device_set_value` after calling `xil_create_from_device`.

Certain attributes, such as `DEVICE_NAME`, must be set with the first method. The "Device Attribute Descriptions" section documents these attributes.

In the following example, the long paragraph attempts to deliver too much information. Almost every sentence in the example conceals a smaller bit of information, which in turn conceals an even smaller bit of information, and so on. Even the reader who understands all of the technical language faces the chore of sorting and retrieving all the information.

Incorrect:

With Gizmo, users of standard mail programs, such as a window-based mail tool, can transparently exchange electronic messages with users of private or public mail systems that conform to X.400 and ISO protocols. Users can reach this broader community without affecting their current electronic-mail routines. Gizmo is both a gateway and a message relay (message transfer agent, MTA, in CCITT terminology). The gateway translates standard mail messages conforming to DoD Simple Mail Transfer Protocol (SMTP) specifications to and from the format specified by X.400. The MTA provides full message analysis and routing. Gizmo builds the Gizmo OSI foundation for messaging over a local area network, and Gizmo OSI combined with Gizmo X.25 for use over packet-switched data networks.

The following rewrite divides the example into several paragraphs and divides some sentences into two sentences.

Correct:

Gizmo enables users of standard mail programs to exchange electronic messages with users of other mail systems. Gizmo works with systems that conform to X.400 and ISO protocols.

With Gizmo, users of mail programs, such as a window-based mail tool, can reach this broader community without affecting their current electronic-mail routines.

Gizmo is both a *gateway* and a message relay. In CCITT terminology, the relay is called a *message transfer agent (MTA)*.

- The gateway translates mail messages that conform to DoD Simple Mail Transfer Protocol (SMTP) specifications. The messages are translated to and from the format specified by X.400.
- The MTA provides full message analysis and routing.

Gizmo builds the Gizmo OSI foundation for messaging over a local area network. Gizmo also combines the OSI with Gizmo X.25 for use over packet-switched data networks.

Eliminate Unnecessary Material

To determine what material is unnecessary, you need to know not only what readers need, but what they do not need. To reduce word count, follow these guidelines:

- Do not include unnecessary definitions and explanations, or information that is too technical for readers' needs.

  Consider creating a glossary for your document. A glossary condenses text by moving definitions out of the main text.
- Eliminate "nice to know" or overly detailed information.
- Avoid flowery language such as "great" and "amazing," buzzwords, and unsupported claims.
- Eliminate introductory text that merely repeats the content in the headings.
- Avoid writing introductions to figures or tables that repeat the figure or table caption.

  In this example of text that introduces a figure, the figure caption reads as follows: "Figure 8–4 Work Item Statistics Page."

  Incorrect:

  When you click the Statistics tab and then the Find Statistics on Work Items link, the Work Item Statistics page is displayed. Use this page to request a statistics report for all work items for one or all applications that are initiated or modified within a given time period. The following figure shows the Work Item Statistics page.

  Correct:

  When you click the Statistics tab and then the Find Statistics on Work Items link, the Work Item Statistics page is displayed, as shown in Figure 8–4. Use this page to request a statistics report for all work items for one or all applications that are initiated or modified within a given time period.
- Omit redundant figure callouts.

  Eliminate callouts for buttons or other screen features that already have a self-explanatory name. For example, you probably do not need a callout for the Delete or Copy button.
- Do not duplicate instructions about how to use a wizard.

  If the instructions are available in online help, do not repeat them in other online technical documentation.
- If possible, perform informal usability testing to see how readers use the document and whether you can eliminate certain types of content.

  If you have control over document navigation, also perform usability testing to improve its design. A good design helps you avoid wasting time and words writing text that attempts to resolve navigational flaws.

Be Careful When Condensing Text

Do not make text so short that it seems "dumbed down," choppy, abrupt, confusing, or incomplete. You do not want to sacrifice content, readability, quality, tone, or flow.

Most importantly, do not remove content that readers need to complete necessary tasks. Also, if sections can be read in any order, you might need to provide cross-references or repeat some information to provide sufficient context for readers. For example, repeat the spelled-out version of an abbreviation or acronym at the first occurrence in each chapter or topic. For more suggestions, see "Preserving Context in Online Documents" on page 125.

Write for the Reader

As a writer, you research, organize, and communicate information for the reader, who depends on you. In your relationship with the reader, you are the expert. Keep this point in mind when you make decisions about what information to present and how that information addresses the reader's questions.

Make Decisions for the Reader

Often a product provides several different ways to accomplish a single task. You might decide that you owe the reader an explanation of each method. However, remember that the reader is more interested in using the product than in understanding all options. Choose the best method for most of your audience, and tell the reader to use that method.

After you commit to the best course of action, you might explain to the reader that other methods exist. Tell the reader where to find your descriptions of those methods. Also, tell the reader why and in what situations options A, B, and C are useful.

For example, the writer of a user's guide for a DOS application included all possibilities in this text:

> The system then displays the following message:
>
> `Accept the path C:\GIZMO? (y/n)`
>
> Type `y` to accept the default path `C:\GIZMO`, or `n` to designate a different path.

Comments: The writer reveals consequences but no guidelines. The details seem to be there, but the entire passage is ambivalent. The writer does not tell the reader why the choice exists. The reader is left to decide without guidance.

Watch for words that could lead to unguided choices. Avoid ambivalent words and phrases, such as the following:

- It is possible to
- Maybe
- Perhaps
- Either, or
- If you want
- Should, would

When you write these words or phrases, or similar ones, make sure that you are prepared to explain the benefits of the choices.

Rewrite: The writer recognized that the passage was ambivalent because the passage did not guide the reader. The rewritten passage guides the reader with this explanation.

> The system displays the following message:
>
> `Accept the path C:\GIZMO? (y/n)`
>
> If you keep all your applications in a particular directory, or if you want to store `GIZMO` on a different hard disk, type `n` to specify your own path.
>
> If you want to create the default `C:\GIZMO` path, type `y`.

Anticipate the Reader's Questions

One of the most important contributions a writer makes is to anticipate the reader's questions and provide appropriate answers. A writer must anticipate questions about related topics as well and provide cross-references to where those questions are answered. As the subject matter expert, a writer can create a climate of understanding that is far more significant than merely recounting facts about the product.

For example, when you review a procedure in your document with the reader's perspective in mind, ask these questions:

- What assumptions have I made about what the reader knows?
- Do steps follow in a logical sequence? Are there any gaps in the instructions?
- Are even the simplest words used precisely? For example, did I write "any" when I meant "all"?
- Did I define all technical terms?
- Have I incorrectly put conceptual and explanatory material within steps, rather than in paragraph text?
- Did I structure each step so that the condition is stated before the action?

  Write, for example, "If the card's I/O address conflicts with another device, remove the card and change the I/O address according to the manufacturer's instructions." Do not write "Remove the card and change the I/O address according to the manufacturer's instructions if the card's I/O address conflicts with another device." State conditions before actions unless this practice needlessly restricts you.

Anticipate Questions

In the following example, the writer of a tutorial clearly explains the function of the clipboard. However, the writer realizes that the explanation might lead to a question: "What happens if I cut or copy another selection?" By anticipating this question, the writer is ready to answer the question in a Note.

> When you cut or copy text, the text is put aside for you on the clipboard, a temporary text storage facility. When text is on the clipboard, you have the option of pasting the text back into the file in any location you choose.
>
> ---
>
> **Note** – As soon as you cut or copy text again, the most recently cut or copied selection replaces the text previously on the clipboard.
>
> ---

Use Cross-References to Address Anticipated Questions

A lack of cross-references can cause the reader great frustration. Often the reader of technical manuals skips important sections. You can presume, therefore, that the reader has not read anything in the book other than the current topic of the current paragraph.

For example, the reader has a question about the file system hierarchy. The reader opens the manual, finds the topic in the table of contents, turns to the page, and reads:

> As mentioned above, the file system directory hierarchy is a part of the "landscape" that you want to become familiar with.

Comments: No reference to either the hierarchy or the "landscape" appears above this sentence. The sentence dooms the reader who has not read everything. The writer presumes that the reader has carefully read everything before the statement that the hierarchy or "landscape" has already been mentioned. Consequently, the reader who goes directly to this section must search for the information.

When using cross-references, do not use the words "above" and "below" to refer to items that are *literally* above and below. Remember that something that appears "above" in today's draft could be "on the previous page" in tomorrow's draft, and these terms are even more problematic in online documents. Using the words "next," "following," "previous," and "preceding" is acceptable if the item referenced is nearby.

Do not write phrases like "As stated in a previous chapter." This reference is too far away to use "previous." Chapters have numbers and names, sections have names, and pages have numbers. Find the location of the information. Cite the location in a specific cross-reference. If the information is not too long or complex to repeat, repeat it.

Rewrite: Cite the specific location of the information:

> As explained in detail in "Issues" on page 2, the file system directory hierarchy is a part of your computer's "landscape."

Avoid Style That Could Offend the Reader

At times, stylistic considerations must go beyond issues of preference. You must also be aware of writing style that could offend the reader. Though offending the reader is not your intention, the use of humor and sexist linguistic conventions can offend. Humor and, especially, sexism are inappropriate in technical writing.

You can also offend the reader by being unintentionally condescending. In keeping your message clear, do not mistake economy of expression for simplicity. Do not underestimate the technical sophistication of the reader.

Avoid Humor

A great temptation for writers of computer documentation is to inject a note of levity into the text. Resist this temptation. Even genuinely humorous commentary is a distraction and becomes annoying on subsequent readings. Likewise, humor that descends into user-friendly chumminess never works. A sympathetic reader might forgive you for trying to "lighten up" the text, but another reader might resent a chummy tone.

For example, this humor was injected into a tutorial:

> You can use a mouse (one without fur) with the window system of your computer.

Comments: The phrase "one without fur" detracts from the content of the sentence and distracts the reader. The goal of the sentence was to tell the reader that the mouse is related to the window system. The apparent goal of the humor was to reassure the reader that the mouse is "friendly."

Rewrite: The following revision pursues those goals in a direct, conversational tone:

> The mouse is a versatile tool that you use with the window system of your computer.

Humor is difficult, if not impossible, to translate successfully. Humor is usually cultural. What might be funny to someone living in one country could be offensive to readers in other countries.

Avoid Sexist Language

Regarding the issue of sexism in language, appearances count.

In many cultures, language has developed so that "men" often refers to "men and women," and "he," "him," and "his" are regarded as gender–neutral words. In decades past, this sentence might have been perfectly acceptable:

> Ask your system administrator for his advice.

Today, this usage of "he" and "his" is far less acceptable. These pronouns assume too much about the gender of an individual. Writers who defend the use of such pronouns must contemplate the following: Many readers could interpret a writer's intentions negatively and could consciously or subconsciously reject the work.

Use Acceptable Methods to Achieve Common Gender

To achieve common gender, you could do the following:

- Use plural antecedents and plural pronouns as often as possible.

  Awkward: Tell each user to shut down his machine.
  Better: Tell the users to shut down their machines.

- Eliminate gender-specific pronouns as much as possible when you are writing in the third person.

  Awkward: Ask your system administrator for his advice.
  Better: Ask your system administrator for advice.

- Use the word "you."

  Awkward: The user should place the file in his home directory.
  Better: Place the file in your home directory.

- Use imperative verbs.

  Awkward: If the user decides he wants to change the settings, he should follow these steps.

  Better: To change the settings, follow these steps.

- Instead of using a personal pronoun, repeat its antecedent when doing so does not sound unpleasant or unnatural.

  Awkward:

  If a system administrator installed the software, wait until he can help you.

  Better:

  If a system administrator installed the software, wait until the system administrator can help you.

Use the following suggestions carefully:

- Give names to "third persons."

  This technique does not work for all types of documentation, but this technique can be effective in a tutorial or other type of user's guide. Consider using names, male or female, to humanize your writing and eliminate the "he" or "she" clumsiness.

For example, if you want to tell the reader how to copy a file from someone else's directory, try this approach:

> Before you can copy a file from someone else's directory, Sally Smith's directory for example, you need permission. Ask Smith to set her file permissions to grant you access. After she has changed permissions, you can copy the file.

- Create your own techniques.

  Keep in mind that the writing should sound natural, be taken literally, and inform.

Avoid Unacceptable Methods to Achieve Common Gender

Eliminating the appearance of sexism by writing poorly, ungrammatically, or self-consciously is not a good solution. Keep the following guidelines in mind:

- Never write "s/he."
- Use "their" with a plural antecedent.

  For example, "ask your system administrator for their advice" is incorrect.
- Try to avoid "his or her," which is grammatically correct but awkward.
- Even in pursuit of the goal of eliminating perceived sexism, never dehumanize people with the pronoun "it."

Do Not Talk Down to the Reader

A "naive reader" is not unsophisticated. Treat the reader as a peer.

Unintentionally, this sentence is condescending:

> Don't be afraid to play with the computer. The computer won't bite you.

Comments: The sentence belittles the reader's anxiety. In an effort to reassure the naive user, the writer seems to belittle the reader's genuine fear of doing something wrong. Of course, the computer will not "bite" the user, and of course the reader can see the writer's point. But the writer makes this point as though writing for a juvenile. A reader might find the tone condescending and insulting.

Remember that the reader might be naive only in relation to the particular computer technology you are documenting. Credit even the least experienced computer user with intelligence and life experience. Show respect for this person when you write.

Rewrite: Get directly to the point, and be positive rather than negative.

> Experimenting with your computer is a great way to learn, and you can quickly undo almost any error.

Common Writing Problems to Avoid

This section identifies words, phrases, constructions, and practices that often lead to abstract or unclear meaning, disjointed cadence, or unnatural and improper language usage.

Anthropomorphisms

Anthropomorphisms attribute human motivation, characteristics, or behavior to inanimate objects. Avoid using anthropomorphisms in technical writing because they can be ambiguous and can confuse readers. However, if an anthropomorphism is an industry standard, you do not have to avoid using it.

Follow these guidelines:

- Use an anthropomorphism if it is an industry standard.

  For example, the verb "listen" is commonly used in the industry in the following way: The system listens to a network to determine when the network is free.

  The attribution of awareness to software has become common, for example, "cluster-aware software," "network-aware software," and "upgrade-aware software." If these terms must be used, define them at first use.

  Many other anthropomorphisms exist that are industry standards, such as "accept," "calculate," "deny," "detect," "interact," "interpret," "refuse," "read," and "write." Whether the use of an anthropomorphism is an industry standard frequently depends on the context.

  For example, the following sentences contain the anthropomorphism "refuse," but only the second sentence shows a correct use:

  Incorrect: The server refuses to boot. (not an industry standard)
  Correct: The network refuses unauthorized connections. (an industry standard)

- If an anthropomorphism is not an industry standard, try to rewrite it by removing or replacing words.

  The following table of anthropomorphic verbs and suggested alternatives is not meant to be comprehensive. It does not include all anthropomorphic terms that are problematic.

  Keep in mind that different parts of speech might be used to anthropomorphize technology. For example, the verb "know," the noun "knowledge," the adjective "knowledgeable," and the adverb "knowingly" might be considered anthropomorphisms if they are used to describe an inanimate object.

  Also, remember that these terms are only anthropomorphisms when they are used with inanimate objects. Therefore, wording such as, "If you want to disable..." is perfectly acceptable.

TABLE 3–1 Anthropomorphisms to Avoid

Anthropomorphism	Suggested Alternatives	Examples
Be interested in	Access, check	Applications that use the service to communicate might access (not "be interested in") the same file or data.
Know	Record, store, detect	The servlet records (not "knows") when a back-end data source was last modified.
Look at	Check, search	The following command searches (not "looks at") the listing to find the `.log` file.
Need	Require	The switch requires (not "needs") one port.
Remember	Store, maintain, save, retain	The software saves (not "remembers") your security profile and activates it the next time you log in.
See	Check, calculate	The system checks (not "sees") whether the physical resources to create the memory segment are available.
Think	Detect, calculate (or reword)	**Incorrect:** If the installation program thinks that the proxy server is installed, you cannot reinstall the proxy server. **Correct:** If the installation program detects the presence of the proxy server, you cannot reinstall the proxy server.
Understand	Interpret, process, handle	The C preprocessor cannot interpret (not "does not understand") Fortran syntax.
Want	Reword	Any remote SDK application that is configured (not "wants") to provide failover protection can use this property.

- Rewrite anthropomorphisms that refer to a company as "we" or "they."

  A company is an "it," not a "we." Another company is not a "they." A company does not recommend, hope, or advise. Instead of saying "We recommend that you remove the cover first," tell the reader "Remove the cover first." Or start with the benefit of the behavior. Instead of saying "It is recommended to use the application's class name as the volume name," say "To ensure that the volume name is unique and meaningful, use the application's class name as the volume name."

- If an anthropomorphism cannot be rewritten without changing the meaning, use a simile with the anthropomorphism or put quotation marks around it.

  Because of translation concerns, sparingly use the technique of putting quotation marks around an anthropomorphic term.

  The following example uses a simile ("...as though...") to avoid directly attributing the human quality of anticipation to the program:

  > The program processes a procedure call as though it had anticipated that call.

The following example demonstrates the ironic use of an anthropomorphism set off within quotation marks:

> The software "takes a snapshot" of the file system.

- Do not ascribe machine qualities to humans.

Use the prepositions "in" and "on" correctly. Do not use "in" when it can be replaced with "use." Do not describe a person as being "on" a server. However, a system can be on a server.

Incorrect: In a text editor, remove the second line of code.

Correct: Using a text editor, remove the second line of code.

Incorrect:

If you are on the server when it fails, you must log in again when the system is operational to retrieve your data.

Correct:

If you are using the system when the server fails, you must log in again when the system is operational to retrieve your data.

Do not tell readers to move around in a file. Instead, use the verb "move" to discuss the pointer position.

Incorrect: Move to the last character in the file.
Correct: Move the pointer to the last character in the file.

Idioms and Colloquialisms

Nonnative English speakers often misunderstand idioms and nonstandard colloquialisms because they interpret these terms literally. Avoid the terms in the following table. In addition to the terms listed in the table, avoid other idioms and colloquialisms that cannot be easily understood out of context.

The suggested alternatives listed in the table are not definitive solutions that will work in all cases. The appropriate alternative depends on the surrounding context, and your best alternative might not appear in this table. Strive to find an alternative that is concise and precise in meaning.

TABLE 3–2 Idioms and Colloquialisms to Avoid

Idiom or Colloquialism	Suggested Alternatives	Examples
At will	As necessary, as you choose	This infrastructure enables you to access the dictionaries from one application or API, mixing and matching them as you choose (not "at will").
Bog down	Overload, slow down, strain, degrade (as in system performance)	Longer timeouts can consume valuable proxy resources and overload (not "bog down") the server.
Building block	Foundation, fundamental, basis, base (or reword)	Knowledge of grammar is a foundation (not "building block") for good writing. **Incorrect:** The disk array can serve as a stand-alone storage unit or as a building block, interconnected with other disk arrays. **Correct:** The disk array can serve as a stand-alone storage unit, or the disk array can be interconnected with other disk arrays.
Carve out	Create, isolate (or reword)	**Incorrect:** To define a nonrectangular surface, specify a set of trimming loops to carve out portions of the rectangular region. **Correct:** Specify a set of trimming loops to isolate a nonrectangular surface within the rectangular region.
Fine-tune	Refine, customize, adjust (or reword)	**Incorrect:** You can fine-tune your system to improve performance. **Correct:** You can refine your system configuration to improve performance.
Geared to, geared toward	Designed for, appropriate for	This application is a standards-based PlirgPak web application framework that is designed for (not "geared toward") enterprise web application development.
Get a feel for	Become familiar with	To become familiar with (not "get a feel for") the user interface, choose the Tour option on the Help menu.
Get by	Reword	**Incorrect:** You can get by without making these changes. **Correct:** You do not need to make these changes.
Hands-off	Automatic (or reword)	**Incorrect:** This task is hands-off. **Correct:** This task does not require manual intervention.
Hinges on	Is determined by, is influenced by, is contingent on, is dependent on, depends on	The decision to set up administrative subdivisions for your network depends on (not "hinges on") several factors.
In concert	In agreement, at the same time (or omit)	The Plirg Directory Interface works with (not "in concert" with) other Plirg technologies to organize and locate components in a distributed computing environment.

TABLE 3–2 Idioms and Colloquialisms to Avoid *(Continued)*

Idiom or Colloquialism	Suggested Alternatives	Examples
In keeping with	In conformance with, conforms to	Ensure that your programming techniques conform to (not "are in keeping with") published guidelines.
In light of	Because of, due to, as a result of	As a result of (not "in light of") these bug fixes, you must recompile your programs.
Keep at your fingertips	For quick retrieval (or reword)	Keep these instructions nearby (not "at your fingertips").
Lends itself	Is amenable to, accommodates, helps (or reword)	This topic is amenable to (not "lends itself" to) discussion. **Incorrect:** The interface lends itself to usability. **Correct:** The interface helps make the system more usable.
On the fly (adv), on-the-fly (a)	Dynamically (for adverb form), dynamic (for adjective form) (or reword)	The HTML code is produced dynamically (not "on the fly") by the server when a particular page is requested.
On the horizon	Avoid wording that makes claims about the future development of products.	
Over the wire	Over (or through) the Internet (or omit)	**Incorrect:** If a remote broker has two identical subscriptions for the same topic destination, the message is sent over the wire only once. **Correct:** If a remote broker has two identical subscriptions for the same topic destination, the message is sent only once. (Term omitted)
Set aside	Allocate, defer, reserve, place elsewhere (or reword)	If you plan to install a single language, allocate (not "set aside") 0.7 Gbytes of additional disk space for the language. Defer (not "set aside") any network upgrades until you have completed a bandwidth capacity analysis. An `fdisk` partition is a section of the disk that is reserved (not "set aside") for a particular operating system. **Incorrect:** Set your tools aside while you remove the housing. **Correct:** Place your tools on a grounded mat while you remove the housing.
Take stock of	Carefully consider or evaluate	Evaluate (not "take stock of") your hardware assets.
To that effect	With that meaning (or reword)	**Incorrect:** Add something to the effect that the web site is password protected. **Correct:** Add a statement indicating that the web site is password protected.

Phrasal Verbs

A *phrasal verb* is a verb followed by a preposition or an adverb. The combination either creates a meaning different from the original verb or is redundant. In most cases, you can replace a phrasal verb with a one-word verb. However, the following phrasal verbs are acceptable when they are used in the appropriate context:

- back up
- check in
- check out
- follow up
- lock up
- look up
- power down
- power off
- power on
- power up
- set up
- shut down
- time out
- turn off
- turn on
- work around
- wrap around

The following table lists phrasal verbs that can easily be replaced and their alternatives.

TABLE 3–3 Phrasal Verbs to Avoid and Their Alternatives

Phrasal Verb	Alternative
Add on	Add
Boot up	Boot
Bring down	Write "cause the system to fail," "shut down the system," or "power off the system," depending on the meaning.
Bring up	Write "power up the system," "start the system," "turn on the machine," or "turn on the power to the system," or other text, depending on the meaning.
Call out	Specify
Connect together	Connect
Dial up	Dial
Figure out	Determine
Fill in	Enter, type
Fill out	Complete
Get rid of	Remove
Group together	Group
Have to	Must
Leave out	Omit
Make use of	Use
Pops up	Appears, is displayed
Print out	Print
Shut off	Power off
Start up	Start
Tie up	Consume

Commands as Verbs

Command names are only names. Command names are never verbs.

> **Incorrect:** First `cd` to the new directory.
> **Correct:** First change to the new directory by using the `cd` command.

Redundancies

You create redundancies when you fail to consider the literal meanings of the words that you choose. Some common examples of redundancies and alternatives are listed in the following table.

TABLE 3–4 Common Redundancies and Their Alternatives

Redundancy	Alternative
Accidental mistake	Mistake
Add additional	Add
Already exists	Exists
At this point in time	At this point, at this time, now, currently
Basic fundamentals	Fundamentals
Check to be sure	Check, ensure, make sure
Close proximity	Close, near, nearby
Contains a listing of	Lists
Create a new	Create
Edit an existing	Edit
Existing conditions	Conditions
First create	Create
In conjunction with	With
Necessary prerequisites	Prerequisites
Specific requirements	Requirements
Still pending	Pending
Whether or not	Whether, if

Test the usefulness of each modifier you choose. If the modifier does not help the construction, for example, if it does not amplify, clarify, or intensify the meaning of the word that it modifies, do not use it. If the modifier's meaning is equal to the meaning of the word that it modifies, or if the phrase or clause is a virtual restatement of a point previously made in the sentence, do not use it.

Well-used intensives, however, often add emphasis to the sentence. For example, "turn on your system for the very first time" might appear redundant at first glance. Something cannot be any more "first" than first. But the intensive, "very," makes clear that the ensuing description happens only once in the system's life. Use this technique sparingly.

◆ ◆ ◆ CHAPTER 4

Structuring Information

This chapter provides information about organizing and presenting your information so readers can find information quickly. The chapter contains the following sections:

Organization Schemes

This section examines different ways to structure information. These structures are not mutually exclusive. You can use these high-level and low-level organizational strategies together in the same document. Some ways to structure a document are as follows:

- **Hierarchy**. Use for related portions of information that are linked together. This structure organizes content at the *section* level within a chapter or stand-alone topic. For more information, see "Organize by Hierarchy" on page 113.
- **Inverted pyramid**. Use for text within individual sections. This structure organizes content at the *paragraph* level and *sentence* level. For more information, see "Organize by Inverted Pyramid" on page 114.
- **Table**. Use for information that you can structure uniformly in columns and rows. For more information, see "Organize by Table" on page 115.
- **Flow diagram**. Use for high-level overviews to show how conceptual information or groups of tasks or procedures are related. For more information, see "Organize by Flow Diagram" on page 116.
- **Task map**. Use for task-based information to give users quick access to step-by-step instructions. For more information, see "Organize by Task Map" on page 118.

Organize by Hierarchy

In a document organized by hierarchy, high-level generalities and overviews offer a preview of what lies below. Levels within hierarchies can be based on importance, frequency of use, or complexity.

Be careful not to create too many hierarchical levels that bury important information. You must balance the need to divide text into short, self-contained sections against the risk of creating valueless intermediate levels.

In a technical manual that is organized by hierarchy, the top three levels serve as a table of contents from which readers can navigate to various sections. Within the text of the document, readers can navigate to related material, both in and out of the document, through cross-references.

In the following example of an online document, a plus sign indicates that a section contains hidden subsections.

EXAMPLE 4–1 Organizing Information by Hierarchy

iPlirg 5769 Reference Manual
+ Preface
+ Back Panel Connectors
Media Independent Interface Connector
    + Twisted-Pair Ethernet Connector
    + SCSI Connector
    Audio Ports
    Audio Specifications
+ Modem Setup Specifications
Identifying Jumpers
    Flash PROM Jumpers
    Serial Port Jumpers
+ System Specifications

Organize by Inverted Pyramid

When organizing by inverted pyramid, you place the conclusion and a short summary of the main ideas at the beginning of the section. The details follow in decreasing order of importance. Readers can digest the main points even if they stop reading before reaching the end of the section.

The inverted pyramid structure, typically used in newspaper writing, is also appropriate for long narrative text in technical documents. Use this structure to organize paragraphs and sentences within a section of narrative text.

To create an inverted pyramid structure, follow these guidelines:

- Use clear, meaningful headings or lists at the beginning of a section.
- Create separate paragraphs or sections to emphasize important points.
- Do not bury your main point in the middle of a paragraph or section.

In the following example, the text begins by defining "device driver optimization." The text then provides links to sections that describe three types of optimization guidelines: general, transmit, and receive. Details appear later in three separate subsections.

EXAMPLE 4–2 Organizing Information by Inverted Pyramid

<Head1>Optimization Guidelines for Network Device Drivers
Optimization in a device driver means that the software interacts directly with a hardware device and mainly fields asynchronous interrupt events.

To assist you in the design of network device drivers in the Solaris kernel, this paper provides three types of optimization guidelines:
- General
- Transmit
- Receive

This document assumes that you have some knowledge of kernel and device driver programming. For a tutorial that covers network device drivers, see `http://www.plirgware.com/net-tutorial/index.html`.

<Head2>General Optimization Guidelines
Some guidelines are universal:
- Use tail calls for stack frame creation.
- Avoid deeply nested automatic variables.
  Compilers can do a better job of register allocation when the scope of automatic variables is limited.
- More guidelines.

<Head2>Transmit Optimization Guidelines
The transmit path requires attention from a driver writer to keep machine performance from being affected under heavy load. To avoid problems, follow these guidelines:
- Queue only when necessary.
- Keep packets in order when queuing.
- More guidelines.

<Head2>Receive Optimization Guidelines
Receive is the most difficult path to tune because many factors are outside the driver's control. For example, packet size and arrival are determined by the remote machine, and the behavior of the protocol stack plays a much greater role than it does on transmit. Follow these guidelines:
- Copy small packets.
- If you must drop a packet, do not tail drop.
- More guidelines.

Organize by Table

Tables are an effective way to compare facts and enhance reader comprehension. However, tables with many columns and rows can be difficult to read and display online. Make tables short and simple to improve readability and reduce online loading time.

If readers are likely to print an online table for future reference, print the table and ensure that it does not run off the printed page.

To simplify tables, follow these guidelines:

- Limit the number of columns and rows.

  For example, you can divide long, wide tables into several short tables.
- Use few words within table cells.

  Use phrases rather than sentences to eliminate unnecessary text. Use abbreviations as necessary. However, do not omit articles where they are needed.
- Consider using a table instead of a bulleted list for long lists with repeating elements.

For more guidelines on constructing tables, see "Tables" on page 70.

In this example, a table is used to organize the locations of various utilities and files for the Solaris and Linux platforms.

EXAMPLE 4–3 Organizing Information by Table

Installation Status	Procedure	For More Information
Before software is installed	1. Deselect the `QInstall` package. 2. Create and run a `finish` script that creates a file named `plirginst`.	__*PlirgSoft Advanced Installation Guide*__
After software is installed	1. Use the `instremove` command to remove the `QInstall` package. 2. Move the `plirginst` file to the `utilities` directory.	__Chapter 12, Software Administration (Tasks) __*PlirgSoft Advanced Installation Guide*__

Organize by Flow Diagram

Use flow diagrams to direct readers through related topics. For instance, you can use flow diagrams to lead readers through a complex task consisting of multiple procedures. Cross-references can take readers to step-by-step instructions for each procedure.

Tip – When authoring primarily for online delivery, consider making flow diagrams clickable.

If the product you are documenting lends itself to a high-level overview, you can use a flow diagram to show how groups of procedures are related. This strategy can help give readers the "big picture" of the product.

Note – Create flow diagrams only if the complexity of your material makes the effort worthwhile, and if doing so is feasible with your authoring tool. Contact your illustrator or production staff for help.

In the following example, a flow diagram helps users troubleshoot the iPlirg appliance in reference to a PROM report. This flow diagram was created using a table. Note that the HTML version of this style guide shows tables with horizontal and vertical lines, regardless of how the table is designed. The printed version accurately reflects how the different tables should appear.

EXAMPLE 4–4 Organizing Information by Flow Diagram

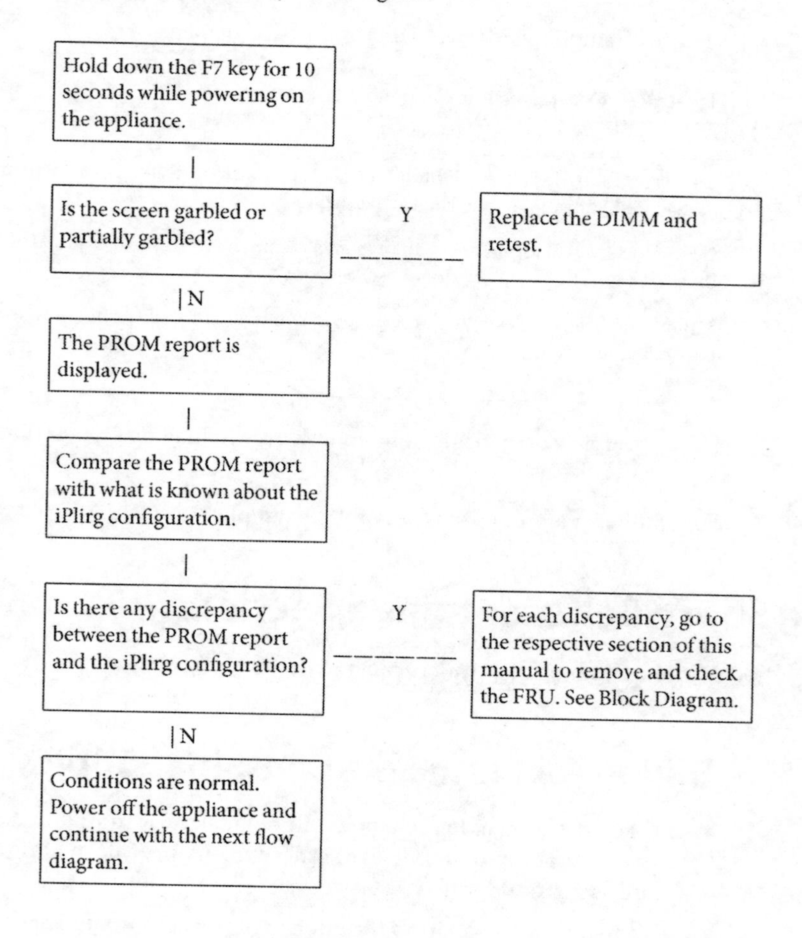

Organize by Task Map

Task maps organize task-based information for users who need fast access to procedures. You can organize individual tasks into a task map that lists either of the following:

- Required, sequential procedures related to a complicated task
- Optional procedures related to a particular topic, such as "Maintaining a Printer"

Cross-references to overview information directly before or after each task map enable readers to view overview material only if they want to do so.

A task analysis can help you determine which tasks are best grouped. Ideally, include on your analysis team users who perform these tasks.

See also "Using Task Maps to Organize Tasks" on page 143.

EXAMPLE 4–5 Organizing Information by Task Map

Task	Description
1. How to Configure a File System for Quotas	Edit the `/etc/vfstab` file to activate quotas each time the file system is mounted. Then, create a quotas file.
2. How to Set Up Quotas for a User	Use the `edquota` command to create disk quotas and inode quotas for a single-user account.
3. (Optional) How to Set Up Quotas for Multiple Users	Use the `edquota` command to apply prototype quotas to other user accounts.

For an overview of managing quotas tasks, see Managing Quotas (Overview).

For another example of a task map, see Example 7–2.

Writing Short, Self-Contained Sections

Write sections so that readers can find needed information quickly.

Divide Text Into Self-Contained Topics

- Explain the content in each topic without making assumptions about previous sections. Cross-references to overviews or other explanatory information can help readers who need more background.
- Consider the complexity of the information and your audience's knowledge and experience when determining how much information to put in a topic.

- Make each topic clearly focused and coherent so that it answers one question about one subject for one purpose.

  Try writing a single question that the topic is meant to answer. Then, judge whether the content fully answers that question.
- Repeat contextual information within each topic when needed.

  Contextual information helps orient readers so that they know how the current topic fits within the larger document structure. See "Preserving Context in Online Documents" on page 125.
- Include subheadings to make text more scannable for readers.

  Subheadings also help readers more easily locate specific information in a table of contents.

Include Transitions

Use transitions between and within paragraphs to show how ideas relate to each other.

After headings, however, do not use transitional text that merely repeats the content in headings. This type of transitional text is unnecessary and redundant.

The text in the incorrect example lacks transitions and unnecessarily repeats the content in the heading. The text in the correct example includes transitional words and phrases, which are emphasized for this example. Also, the introductory sentence is no longer redundant.

EXAMPLE 4–6 Including Transitions

Incorrect:

<Head2>How the Plirg Secure Environment Protects Against Intruders

The Plirg Secure environment protects against intruders by providing accounts that require user names with passwords. They can be created by users or be system-generated, according to your site's security policy. You can require passwords to be changed regularly.

Users can work within their approved label range only, which limits the information they can access. Additional passwords are required for certain administrative tasks. This limits the damage that can be done by an intruder who guesses the superuser password.

Correct:

<Head2>How the Plirg Secure Environment Protects Against Intruders

Plirg Secure *protection* includes accounts that require user names with passwords. *Passwords* can be created by users or can be system-generated, according to your site's security policy. You can *also* require passwords to be changed regularly.

In addition, users can work within their approved label range only, which limits the information they can access. Additional passwords are required for certain administrative tasks. *This requirement* limits the damage that can be done by an intruder who guesses the superuser password.

Online Writing Style

Online documents have characteristics that distinguish them from their printed counterparts. For example, online documents have the capability of linking to related information. However, they also have screens that are less easily read than printed pages.

Writers must follow guidelines for online writing style in addition to style guidelines that apply to all forms of technical writing.

This chapter discusses the following topics:

About These Guidelines

The guidelines in this chapter focus on online text that is intended to be read in a web browser . This chapter does not discuss software tools, web browsers, screen design, graphics, or document delivery methods. Underscores before and after link text show where to place links in the text.

The online writing guidelines in this chapter do not apply to PDF or PostScript documents.

This chapter uses the word "page" to refer to the information that is displayed in a scrollable window or dialog box. When "printed" precedes the word "page," the reference is to hard copy.

Note – For single-source document writers: If you encounter a conflict between online and print writing guidelines, follow the guidelines for your primary delivery mode.

Solving Online Writing Problems

Online writing presents special challenges for writers. The following table summarizes the problems that are unique to online writing and presents possible solutions.

TABLE 5–1 Online Writing Problems and Solutions

Problem	Solution
Online readers cannot easily envision the size or complexity of an online document or the relationship among various topics.	Create an effective document structure that is obvious and easy to navigate. See "Creating an Effective Online Document Structure" on page 122.
People do not like reading online text and become frustrated when they cannot quickly find what they want.	Write short, self-contained topics and construct simple, scannable text. Both strategies help reduce information overload and make it easier for readers to find quick answers to their questions. See "Writing Online Topics" on page 124 and "Constructing Scannable Text in Online Documents" on page 124.
Text is more difficult to read online.	Construct simple, scannable text. See "Constructing Scannable Text in Online Documents" on page 124.
Most readers go directly to a topic to find information. Because they have no context from earlier reading, they can easily become disoriented.	Preserve context to keep readers from becoming lost or disoriented in a document. See "Preserving Context in Online Documents" on page 125.
Links can be annoying, especially when the content and value of the material located at the link destination is unclear.	Construct effective links that enable readers to make their own decisions about how to access information. Avoid distracting or disorienting readers by overlinking or by providing insufficient context for links. For more information about links, see Chapter 6, "Constructing Links."

Creating an Effective Online Document Structure

Readers cannot easily visualize the size or complexity of an online document because they view only one window of text at a time. Both the absence of hard copy to reveal the document's general dimensions and the small window size also create challenges for readers. These challenges combine to make it difficult for readers to see how various topics are related.

Readers who must retrace their steps or move forward with no clear idea of where they are going are easily frustrated. To reduce the risk of reader disorientation, help readers visualize the structure of your online document.

To create an effective document structure, do the following:

- Determine how much control to give readers.

  When planning your online document, decide how much control readers should have over the sequence of topics or the paths that they can follow. Consider the purpose of your document, the scope and complexity of your technical content, and the skill level of your audience with online media.

- Provide links that anticipate readers' needs.

  Provide links that are embedded in the text and that point to other relevant content in the document. Also provide links in a jump list. See "Writing Jump Lists" on page 69. For guidelines on writing link text, see Chapter 6, "Constructing Links."

- Use an easy-to-follow, meaningful structure.

  If your document requires many clicks to find the information readers seek, rethink its structure.

Note – Do not describe how to navigate in your online document unless it has truly unusual features or you anticipate that many inexperienced online readers will use your document.

The following list suggests ways to structure an online document. These structures are not mutually exclusive. You can use these high-level and low-level organizational strategies together in the same document.

- **Hierarchy**. Use for related portions of information that are linked together. This structure organizes content at the *section* level within a chapter or stand-alone topic. For more information, see "Organize by Hierarchy" on page 113.
- **Inverted pyramid**. Use for text within individual sections. This structure organizes content at the *paragraph* level and *sentence* level. For more information, see "Organize by Inverted Pyramid" on page 114.
- **Table**. Use for information that you can structure uniformly in columns and rows. For more information, see "Organize by Table" on page 115.
- **Flow diagram**. Use for high-level overviews to show how conceptual information or groups of tasks or procedures are related. For more information, see "Organize by Flow Diagram" on page 116.
- **Task map**. Use for task-based information to give users quick access to step-by-step instructions. For more information, see "Organize by Task Map" on page 118.

For more information about structuring your document, see Chapter 4, "Structuring Information."

Writing Online Topics

Write topics that minimize screen readability problems and that help readers find needed information quickly. Follow these guidelines:

- Divide text into self-contained, linked topics.

  For more information, see "Divide Text Into Self-Contained Topics" on page 118.
- Keep topics short.

  Online topics can be quite short. However, just one or two short sentences probably do not justify splitting off the text into a separate topic.
- Focus on the structure of the content, not on how formatting renders online.
- Include transitions within each topic.

  For more information, see "Include Transitions" on page 119.

Constructing Scannable Text in Online Documents

Readers expect online text to be succinct and directly relevant. They are also goal-oriented and scan for information rather than read long blocks of text thoroughly. Make online documents scannable to help readers locate information quickly.

To construct scannable text in online documents, do the following:

- Write clearly and simply.

  For more information, see "Write Clearly and Simply" on page 95.
- Keep paragraphs short.

  For more information, see "Keep Paragraphs Short" on page 96.
- Condense text.

  Write concisely and eliminate unnecessary material. However, be cautious about condensing text too much. For more information, see "Be Careful When Condensing Text" on page 99.
- Use links to make text seem shorter.

  Text seems shorter because readers use links to select only the topics of interest. See "Use Links to Make Text Seem Shorter" on page 133.
- Replace text with tables, charts, and figures, when possible.
- Write meaningful headings and subheadings.
- Use bulleted lists and jump lists.

Preserving Context in Online Documents

Online readers create their own transitions as they move through a document. Through traversing links, both associative and navigational, readers choose their own path by deciding where to go next.

This freedom to jump around online has a disadvantage. Sometimes, context is lost and readers can become confused or disoriented. To preserve context, online documents must support nonlinear and incomplete reading. Follow these guidelines:

- Make few assumptions about reading order.
- Offer contextual cues.
- Give the precise location of related information.
- Use pronouns carefully.

  The antecedents of "this" and "it" are often unclear, especially online. Pronouns that are too vague also cause problems for translators. For more information, see "Pronouns" on page 34.

Make Few Assumptions About Reading Order

In an online environment, you usually do not know how readers come to your document or a section of it. They might arrive through a table of contents or a search engine. They might click a cross-reference that leads them to your document. They could just be scanning links.

When you make few assumptions about reading order, you are better prepared to offer readers the necessary contextual cues.

Offer Contextual Cues

Contextual cues help readers understand where the information belongs within the larger structure of the document.

If you do not provide sufficient context, readers have to figure out how the topics are related. You need to help readers make these associations meaningful, for example, by indicating to readers that a certain topic is part of a broader topic.

For example, to indicate that information under a third-level heading is a subsection of a second-level heading, you could write the subsection heading to reflect this relationship. Or, you could include introductory text to this effect.

If sequential order is important, give readers a sense of what comes before and after a topic. For example, if a complex task is logically divided into several short subtasks or procedures, tell readers what comes before and after each procedure. A list of prerequisites before a procedure and a "Where to Go From Here" section after a procedure would suffice here.

Give the Precise Location of Related Information

The words "above" or "below" have no meaning in a document that is not read linearly. Instead, use one of these strategies:

- Use "next," "following," or "previous" if the information is contained in the same topic.
- Name the item.

  For example, instead of using the text "in the following section," you can name the section by using the text "Modifying a Group."
- Link to the information.

For example, you could use the following format in an online, linear document.

EXAMPLE 5–1 Preserving Context in Online Documents

Incorrect:

<Head2>Demo Programs

You can find demo programs in `C:\Demos`. See the README file for detailed information about them.

<Head2>Related Documentation

For more information about this topic, see the Application Builder help volume and the *Application Builder User's Guide*.

Correct:

Previous Topic: Developing an Application

<Head2>Application Builder Demo Programs

You can find the Application Builder demo programs in the `C:\Demos` directory. See the README file for detailed information about these programs.

<Head2>Application Builder Documentation

For more information about Application Builder, see the Application Builder help volume and the *Application Builder User's Guide*.

Next Topic: Designing and Maintaining Portable Applications

Note – Explicitly state Previous Topic and Next Topic only if your document requires this format and the display tool does not automatically generate this information.

◆ ◆ ◆ CHAPTER 6

Constructing Links

A link consists of the *link text* and the *link destination*. Links enable readers to jump to related information with a single click. Links are one of the advantages of online reading because they enable readers to make their own decisions about how to access information.

However, links pose challenges for writers. Too many links clutter text and can distract readers. Too few links mean readers might have difficulty finding the supplemental information that they need when looking at a screen of text, which lacks the context of a printed page.

Note – Depending on your writing tool, you might be unable to implement some guidelines in this chapter.

This chapter discusses the following topics:

About These Guidelines

The guidelines in this chapter focus on online text that is intended to be read in a web browser. This chapter does not discuss software tools, browsers, screen design, graphics, or document delivery methods.

This chapter uses underlines to show where to place links in the text. For example: PlirgSoft Manager uses virtual disks to manage physical disks.

The online writing guidelines in this chapter do not apply to the following types of documents:

- PDF documents
- PostScript documents

This chapter uses the word "page" to refer to the information that is displayed in a scrollable window or dialog box. When "printed" precedes the word "page," the reference is to hard copy.

Where to Place Links

When considering which links to create, anticipate paths that your readers are likely to take. Which associations would benefit them? Let your readers' needs, expectations, and interests guide you. Create links that support good decision making.

When readers must follow a certain sequence through your content, for example, when following a procedure, limit their choices by reducing or eliminating links. At the least, guide readers with additional explanatory text.

Except for such instances, your goal is to provide readers with opportunities, not to order them around. However, do not make these opportunities endless. You must set priorities in your documents and point your readers in relevant directions.

The following list suggests ways to use links in online documents.

- **Tables of contents.** If your writing tool does not generate a clickable table of contents for you, include one as the entry point into your online document. If your writing tool can expand and collapse entries for tables of contents, use this feature.
- **Summaries.** To reduce the amount of information that readers see, briefly summarize content and then link to supporting details. For example, you can link to long examples and to overview, background, reference, detailed, or supplementary information.

  If you organize your document by hierarchy, you can present general overviews that link to increasing levels of details. See "Organize by Hierarchy" on page 113.
- **Lists.** Online readers interpret links more easily in bulleted lists rather than links embedded in paragraphs. A jump list can serve as a short table of contents or a brief summary of the content if it is placed at the beginning of a chapter or main section of a document.

  For more information, see "Provide Links in a List" on page 133.
- **Tables.** Tables with links to related information provide scannable content and quick access to more details. See "Organize by Table" on page 115 and "Tables" on page 70.
- **Cross-references.** Cross-references to other sections, chapters, or manuals typically become links embedded in the text. Consider clustering cross-references to headings or titles under a "See Also," "Related Topics," or similar heading.

  Capitalize and punctuate cross-references by following the rules in "Formatting Cross-References" on page 77.

- **Code examples.** Consider placing long blocks of programming code in a separate text file. Then, link to the text file from your document. This strategy eliminates the formatting cleanup that is often necessary when a user cuts and pastes code from documents displayed in a browser.
- **Glossary terms.** Wherever the prominent use of a glossary term appears, link to the document's glossary if the document has one. In determining the prominent use of the term, consider the reader's likely path through the document. Remember that too many links can distract readers. Balance the desire to provide helpful links to the glossary against the risk of overlinking.
- **Reader comments.** Include an online feedback mechanism on every page of your online document if you have control over this information design feature. Direct comments to the appropriate group, not, for instance, to the webmaster if your document is displayed on the web. Also, let users know if they should expect a response to their comments.
- **Reference lists.** A document such as a white paper can link to a reference list that is contained in a separate page. In print, the reference list would be contained within the body of the paper.
- **Other relevant web sites.** For web documents, links to material on other sites enable you to take advantage of what others have produced. For information about policies for linking to third-party web sites, see "Referencing External Web Sites" on page 194.

General Linking Strategies

Links stand out by virtue of being underlined and displayed in color, so think of them as emphasized words. Because the scanning eye notices only two or three words at a time, emphasized or not, make the link text short but informative.

When constructing links, use these general strategies:

- Create links that anticipate readers' likely paths.

  See "Where to Place Links" on page 128.
- Avoid overlinking.
- Prevent reader disorientation.
- Include links that answer readers' questions.
- Use links to condense the amount of information readers see.
- Provide links in a list when possible.
- Place links at the end of a topic when possible.
- Provide URLs only when needed.
- Test the validity of links.

Avoid Overlinking

Be careful about filling text with distracting links. Too many links dilute the message and can confuse readers with irrelevant digressions. You need to guide readers and filter their choices.

Links also disrupt the narrative flow of text by inviting readers to go elsewhere. Unless that is your goal, link sparingly. Also, remember that too many links can involve too much maintenance and that they are hard to read on a cluttered screen.

The following sections suggest ways to avoid overlinking.

Avoid Interrupting Readers

If you do not want to interrupt the reader at a certain point in the text, do not put a link there. Ask yourself: Is the information at the link destination relevant to the audience, purpose, and content of this document?

Do Not Link the Same Text Repeatedly

- Do not repeat a link wherever the link text occurs.

  Identifying a link once per topic is sufficient.

  Incorrect:

  PlirgSoft uses file folders to organize your files. These file folders can contain only graphic and text files.

  Correct:

  PlirgSoft uses file folders to organize your files. These file folders can contain only graphic and text files.

- Do not link to an online feedback mechanism more than once on a page.
- Avoid making explicit cross-references if they create redundancy.

  Incorrect:

  If you used Wrap To Fit, the Save dialog box includes an additional choice about handling line endings (see Wrap To Fit).

  Correct:

  If you used Wrap To Fit, the Save dialog box includes an additional choice about handling line endings.

Note – If you are generating a printed and an online version of your document from the same source, you might need to make explicit cross-references to accommodate the requirements of the printed version.

Minimize the Number of Clicks

Do not add a link if you can succinctly present the information in the current topic.

In the following incorrect example, readers need to click twice to access the *Application Builder User's Guide*. The first click, click here, takes readers to the list of related documentation. The second click, which is not shown in the example, takes readers to the user's guide. In the correct example, readers need to click only once to access the user's guide.

Incorrect:

For related documentation on Application Builder, click here.

Correct:

For related documentation on Application Builder, see the Application Builder help volume and the *Application Builder User's Guide*.

Limit Internal Linking

- Especially if a page is short, do not link to other destinations on that page.

  Readers typically expect links to take them to another page. If a link takes readers only a few lines down, readers can become disoriented.

- When a table, figure, or example immediately follows a textual reference to it, do not include a link.

  The unnecessary link can disorient and frustrate readers, especially if clicking on the link renders an entirely new browser window or tab that contains only that structure.

  In the textual reference to the table, figure, or example, use the word "following" to indicate the structure's location. For example, you might begin the sentence with the text "The following figure shows."

Prevent Reader Disorientation

Ensure that readers remain fully oriented and in control as they navigate through an online document. Readers are properly oriented when they can identify the content presented, the location of the content within the larger body of information, and the navigational options.

Some techniques to prevent disorientation include the following:

- Reserve link formatting for links only.

  Never use your link formatting convention for any other purpose. Links are usually identified by underlining and a certain color.

- Label navigational icons.

  Label navigational icons if you have such control over them. Some users are visually oriented while others are text oriented. A combination serves all users.

- Provide sufficient context.

  Sufficient context helps readers decide whether to follow the link. Explain what information is located at the link destination and how it relates to the present topic. See "Provide Context in Link Text and Surrounding Text" on page 135.
- Warn readers about unexpected link destinations.

  Let readers know when a link might take them to an unexpected destination. For example, tell readers when clicking on a link does the following:
  - Opens another document
  - Opens another application
  - Takes them to another site
  - Leads to a large file with a long download time
  - Requires them to register or enter a password to access the destination

  If you carefully word a link and its surrounding text, readers know when the link takes them away from your document.

In the following example, links to "VI" and "Emacs" take readers away from the tutorial. The correct version clearly indicates that clicking on either link takes readers to another site.

EXAMPLE 6–1 Preventing Reader Disorientation

Incorrect:

Pico is probably the easiest of the three editors to use. If you are curious about how to use the other editors, however, see the reference cards for VI and Emacs.

Correct:

Pico is probably the easiest of the three editors to use. If you are curious about how to use the other editors, however, see their reference cards:

- VI Reference Card from Dalhousie University
- GNU Emacs Reference Card from `Refcards.com`

Include Links That Answer the Reader's Questions

One reason for providing links is to answer the reader's questions, such as "How do I compress a file?" Ideally, write link text so that it corresponds to the reader's search tasks.

Incorrect:

Bug voting provides another channel for feedback and for establishing priorities. You can cast a vote for the bugs that frustrate you the most. Your votes are viewed by engineering. If you want to vote, click here.

Correct:

Bug voting provides another channel for feedback and for establishing priorities. You can cast a vote for the bugs that frustrate you the most. Your votes are viewed by engineering.

Use Links to Make Text Seem Shorter

Your online document might be lengthy. However, with links, your text can appear shorter to readers without sacrificing depth of content.

- Link to in-depth information.

  Use summaries and link to supporting details. See "Summaries" in "Where to Place Links" on page 128.
- Link to, rather than duplicate, information.

  For example, if a procedure is basic and frequently used as part of other more complex procedures, do not repeat the steps of that basic procedure. Instead, put the steps for the basic procedure in one section and link to it from other places.

  Also, do not duplicate background or overview information. Include this material in one place and link to it from other places, as needed.
- Link to basic information.

  Link to basics so expert users do not stop reading. Novice users can access the material if they are interested.

Note – Avoid dividing long, linear text into multiple pages with Next links, which is known as "electronic page turning." If this method seems like the only way to present the material, consider providing a printer-friendly version for offline reading or a PDF file that readers can download.

Provide Links in a List

- Try to use a list of links instead of links embedded in the text.

  Searching a list of links rather than a paragraph with embedded links reduces the mental processing demands of online reading and link interpretation.

  Incorrect:

  The installation of the application server is affected by the following resource issues: unique network ports, shared directory configuration trees, shared operating environment, and login.

  Correct:

  The installation of the application server is affected by the following resource issues:

  - Unique network ports
  - Shared directory configuration trees
  - Shared operating environment
  - Login
- Carefully craft the link text, and annotate each link, when necessary, to ensure that readers can easily decide which links to follow.

Place Links at the End of a Topic

To encourage readers to stay with a topic, place links at the end of a topic, when possible. Readers might read all or at least some of the content before moving on.

In the following incorrect example, the link text falls in the middle of the paragraph. In the correct example, the link text falls at the end of the topic.

Incorrect:

You can install the application server by using `setup.exe` or `ezSetup.exe`. The ezSetup method provides an easy installation without requiring various inputs. For more information, see Using ezSetup. However, this section covers the `setup.exe` installation options: Express, Typical, and Custom installations. The Typical option is the default.

Correct:

You can install the application server by using `setup.exe` or `ezSetup.exe`. This section covers the `setup.exe` installation options: Express, Typical, and Custom installations. The Typical option is the default.

For an installation that does not require various inputs, use the ezSetup method.

Provide URLs Only When Needed

Use URLs in the following instances:

- When readers are likely to print the page and then later retype a URL on that page in a browser
- When your online document is also available in print
- When your goal is to raise awareness of a site

  Reserve this usage for short URLs only, such as www.plirg.com, because readers are unlikely to remember long URLs.

Some URLs require a slash at the end and some do not. Before publishing your document, test all the URLs to ensure that they work.

Make references to URLs as simple and as direct as possible. For suggestions on how to introduce URLs in text, see "Referencing URLs" on page 273.

Test the Validity of Links

Testing the validity of links ensures that links in your document lead where they promise. Test the validity of both internal and external links.

Guidelines for Crafting Link Text

When crafting link text, follow these guidelines:

- Provide context in link text and surrounding text.
- Weave link text into sentence structure.
- Choose key words or phrases for link text.
- Choose an appropriate length for link text.
- Write scannable link text.
- Make link text conceptually similar to titles or headings of link destinations.
- Do not use quotation marks around link text.

Provide Context in Link Text and Surrounding Text

Readers use links as guideposts when scanning, so take full advantage of them and word your link text accordingly. Effective link text also creates adequate context for readers and reduces the likelihood of readers becoming disoriented.

- Choose your link text carefully, as well as the text that surrounds the link text.
- If possible, supply explanatory text before the descriptive link.

  This additional information helps readers understand where each link leads and why it was chosen.

  Incorrect:

  Appendix F identifies the system components.

  Correct:

  For a list of the Plirg 5769 system components, see Appendix F, Illustrated Parts Breakdown.

  Incorrect:

  The combination of hypertext and the global Internet started a revolution. In this article, published in *Scientific American*, Jon Bosak and Tim Bray tell how a new tool, XML, is poised to finish the job.

  Correct:

  The combination of hypertext and the global Internet started a revolution. Jon Bosak and Tim Bray tell how a new tool, XML, is poised to finish the job. See the *Scientific American* article, XML and the Second-Generation Web.

Weave Link Text Into Sentence Structure

- Make the text of the link meaningful and part of the natural syntax of the sentence.
- Do not make the link text refer to the mechanism of the online display tool, as in "click here" or "go here."

  One exception is when readers need to see the URL of a web document. See "Provide URLs Only When Needed" on page 134.

  Incorrect: Click here for more information about Plirg File Manager.

  Correct: For more information, see Plirg File Manager .

Choose Key Words or Phrases for Link Text

Choose key words or phrases in text that best represent the content of the destination.

Incorrect:

1. How do you set Thunderbird mail to use the POP protocol?
2. What are known problems with Thunderbird mail for Plirg products?
3. Thunderbird Frequently Asked Questions

In the correct example, the wording of the third question in this example has changed. The reworded text removes the long, unbroken line of link text and presents two different link destinations. Readers now have two link choices instead of one.

Correct:

1. How do you set Thunderbird mail to use the POP protocol?
2. What are known problems with Thunderbird mail for Plirg products?
3. Thunderbird Frequently Asked Questions from Mozilla's site `mozillamessaging.com/support`

Choose an Appropriate Length for Link Text

A link that is the length of a complete sentence is too long and difficult to read. One to three words usually works best, as long as those words are context-rich. Using more words is acceptable if they provide helpful context.

Incorrect:

Plirg 5769 System:

- Plirg 5769 Installation Guide (Download)
- Plirg 5769 Maintenance Manual (Download)
- Plirg 5769 Parts List (Download)

Correct:

For the Plirg 5769 system, you can download the following information:

- Installation Guide
- Maintenance Manual
- Parts List

Write Scannable Link Text

Write links as though your reader were scanning only the links on the page. As the eye processes a page, it jumps to the links, so make them self-explanatory and scannable.

Incorrect:

There will be a moderated online chat on the PlirgSoft API for XML Parsing (PAXP) on Tuesday, April 25, at 11:00 a.m. Pacific Daylight Time. The guests are Wallace Gromit, Plirg architect–XML technologies, and XML parser guru Laurel Hardy. To join the chat, go here.

Correct:

Join the online chat on PlirgSoft API for XML Parsing (PAXP) on Tuesday, April 25, at 11:00 a.m. Pacific Daylight Time.

The guests are as follows:

- **Wallace Gromit**, Plirg architect in XML technologies
- **Laurel Hardy**, XML parser guru

Make Link Text Conceptually Similar to Titles or Headings

Check that the text of all or most of the links in a topic are conceptually similar to the title or headings of their associated link destination.

Also, use consistent wording for link text that leads to the same link destination, if possible. For example, if you have a Feedback Form link at the top of your multipage document, do not use the link text feedback page elsewhere in the same document.

Do Not Use Quotation Marks Around Link Text

Quotation marks around link text unnecessarily clutter the text.

Incorrect: See "Knowledge Management Center."

Correct: See Knowledge Management Center.

CHAPTER 7

Writing Tasks, Procedures, and Steps

The purpose of many technical documents is to explain how to use a product to accomplish specific tasks. In such documents, detailed instructions on how to accomplish tasks are often provided in the form of procedures and steps.

This chapter discusses the following topics:

Understanding the Relationship Among Tasks, Procedures, and Steps

This chapter uses these task-related terms as follows:

- **Task.** Specific work that can be performed. A task includes instructions for completing the work. A task also can include an explanation of why the work might be performed, as well as prerequisites and examples.

  A task can be short and simple, even just one action to complete. A task can also be long and complex. A long, complex task might need to be separated into subtasks to make the task easier to understand.

- **Subtask.** A small, short component of a larger task. A subtask might be one action or one set of actions to complete. A subtask can include prerequisites and examples. If a subtask is long and complex, that subtask can also contain subtasks.

  To complete a task, a user might need to complete or choose from multiple subtasks. Some subtasks might be *optional* or *conditional*. Such tasks might not always have to be completed, depending on the user's situation or the desired outcome. Make certain that you clearly identify any optional or conditional subtasks as being alternatives.

- **Procedure.** One step or an ordered set of steps that explains how to accomplish a task or subtask. If a task or subtask is not complex, the task or subtask might be just one action.

  A procedure can be optional or conditional. A procedure also can include prerequisites and examples.

- **Step.** An instruction that explains how to perform a procedure or part of a procedure. A short, simple procedure might require only one step. Two or more steps are ordered and are numbered to show the sequence of actions.

  A step can be optional or conditional. A step also can include prerequisites and examples.

A short task, such as backing up a system, might require performing one simple procedure. In such instances, you might not use all of the guidelines in this chapter.

Developing Task Information

A task is work that is performed for a particular purpose.

Task orientation is useful for all technologies. Task orientation is as important for a developer's guide as it is for a user's guide or an administrator's guide. When identifying tasks, do not become distracted by the interface. The interface, whether it is a browser interface, graphical user interface (GUI), application programming interface, or command-line interface, is the means by which a user accomplishes tasks. The writer's primary focus should be on how to accomplish the tasks, not on how the interface works.

To develop and write task information, follow these guidelines:

- Perform an audience analysis and a user task analysis.
- Provide only necessary task information.
- Organize related, optional, and conditional tasks.
- Use continuous prose for tasks, when appropriate.
- Do not use procedures for command explanations.

Perform an Audience Analysis and a User Task Analysis

Before you start writing tasks, identify the audience and perform a user task analysis.

An *audience analysis* identifies the readers and the readers' skills. From the audience analysis, you can determine the amount of detail to provide in the tasks that you write.

Product managers and engineers can tell you who the audience is for a particular document. The audience might be end users, developers, or operators, for example. Subject matter experts can tell you the audience's level of experience and training.

A *user task analysis* identifies all possible uses of a product or products or what a user does with the product. A user task analysis can help you determine whether to write about one task or to divide the task into subtasks. From the user task analysis, you can determine which tasks to break down into subtasks and which tasks to group together. You can also determine whether tasks and subtasks are required, optional, or conditional. Consult with subject matter experts as needed.

Provide Only Necessary Task Information

When writing a task, provide only the information that is necessary to complete the task. In particular, limit any overview information to information without which the user cannot complete the task.

To provide just the relevant information, consider including the following information in a task:

- An explanation of what the task is
- The reasons why readers need to perform the task
- Prerequisites for performing the task
- Instructions on how to perform the task
- Examples that illustrate how to perform the task

For most tasks, the instructions on how to complete the task are in the form of a procedure. For guidelines on how to write procedures, see "Writing Procedures" on page 149.

Some tasks are not suitable for being explained as a procedure. For example, tasks that are associated with an API are not suitable for being explained as a procedure. For examples of writing a task other than as a procedure, see "Use Continuous Prose for Some Tasks" on page 145.

Including Prerequisites

Include any prerequisites that users must consider before users perform a task. The risk of users performing an action out of sequence is particularly high with online documents because users can enter a task from various points.

For a task that is written as continuous prose, place prerequisites in an introductory paragraph. For a task that is written as procedures, follow these guidelines:

- If the task contains information that users need to know before performing the procedure, include that prerequisite information in an introductory paragraph.

  The following examples of introductory paragraphs contain such information:

  > Before you start the configuration process, collect the required system and network information, as described in the Information Collection Worksheet.
  >
  > Before you begin this procedure, ensure that your system has at least 56 Mbytes of free memory.
  >
  > Before you power off the system, you must halt any I/O between the host systems and the Plirg 5769 system.

- If users must perform a prerequisite step or procedure, make the first step the prerequisite step or a cross-reference to the prerequisite procedure.

  1. If the system is not already shut down, type `shutdown`.

  If users must become superuser or issue other commands at the command line to enable them to perform a task, make that the first step.

  1. Become superuser.

  ```
  $ su
  Password:
  ```

  1. If the Container Manager GUI is not already open, access it as described in "To Start the Container Manager GUI" on page 57.

  Note – Depending on the technical expertise of your audience, you might not need to include the syntax for the `su` command or other basic commands.

- If some readers are novice users, include a cross-reference to a basic procedure.

  1. If a browser is not already running, start a browser.

  See "To Start a Web Browser" on page 8.

Providing Examples

Consider including one or more examples whenever doing so can help readers. Do not provide an example if the task is self-evident.

A command-line example shows the commands and the resulting output. A continuous prose example shows all of the function calls and other code that are required to complete the task.

Note – Most examples in this chapter show command-line information for software procedures.

When providing an example, follow these guidelines:

- If the example requires clarification, include text with the example.
- Use practical data in the example rather than replaceable or variable text.
- If the caption for an example title contains a verb, begin the caption with the gerund form of the verb.
- Keep the example short, showing only the necessary elements.

  If output is lengthy or used only for verification, show just the first lines, last lines, and pertinent intervening lines. Use vertical ellipsis points to indicate any missing lines that readers do not need to see.

For example:

```
$ eject -n
.
.
.
rmdisk0 -> /vol/dev/rdsk/c4t0d0/clik40     (Generic USB storage)
cdrom0 -> /vol/dev/rdsk/c0t6d0/audio_cd    (Generic CD device)
jaz0 -> /vol/dev/rdsk/c3t0d0/jaz1gb        (USB Jaz device)
```

The following example shows how you could provide an example in a procedure.

EXAMPLE 7–1 Installing the Driver Package (Example of Procedure)

After you download the `PLRGdrv.tar.Z` package to the `tmp` directory, install the driver:

```
$ cd tmp
$ uncompress PLRGdrv.tar.Z
$ tar xf PLRGdrv.tar
$ su
Password:
# ./install.drvr
```

Note – Be careful when providing examples of code or screen captures. Make sure that each name that you use for a URL, computer, IP address, and network domain can be made public. See "Protecting Confidential Information in Examples" on page 200 for details.

Organize Related, Optional, and Conditional Tasks

Consider including jump lists, task maps, or flow diagrams in a document to help readers understand the organization of tasks. The tasks can be related, optional, and conditional, as explained here:

- **Jump lists.** Use jump lists if you want to provide cross-references to simple, related tasks.
- **Task maps.** Use task maps if you want to provide a tabular summary of related, optional, and conditional tasks.
- **Flow diagrams.** Use flow diagrams if you want to provide a graphical representation of related, optional, and conditional tasks.

Using Jump Lists to Organize Tasks

A *jump list* is a list of cross-references that serves as a table of contents for a chapter or section. By using a jump list, readers can quickly go to a particular task within the list of related tasks.

See "Writing Jump Lists" on page 69 for information about constructing a jump list.

Using Task Maps to Organize Tasks

A *task map* organizes tasks in tabular format. A task map lists two or more subtasks that relate to an overall task. A task map also points to instructions for completing those tasks. When writing a task map, do not provide step-by-step instructions.

To create a task map, follow these guidelines:

- Provide a brief introduction to the task map, when appropriate.

  You can also cross-reference related overview or supplementary information.
- If you include a Description column in a task map, make sure that entries explain why the task is performed rather than how to perform the task.
- If users are to perform the tasks in sequential order, number the individual tasks.
- Identify any optional and conditional tasks that appear in a task map for sequential tasks.

  For an example of an optional task and a conditional task, see Task 3 and Task 5, respectively, in the example that follows.
- Cross-reference each task to its related instructions.

  The cross-reference is usually to a book, chapter, or section.

The following example shows a typical task map.

EXAMPLE 7–2 Task Map

The following table shows a task map for setting up a custom QuickStart installation.

Task	Description	Instructions
1. Make sure that the system is supported.	Check the hardware documentation to see if the system is supported.	*PlirgSoft Hardware Platform Guide*
2. Make sure that the system has enough disk space for the PlirgSoft software. Also, consider the disk space requirements for any additional third-party software you might want to install.	Determine which software group to install. The disk space must accommodate a minimum of five file systems. (Optional) Check that the disk space for third-party software is adequate.	Chapter 2, "Planning Disk Space"
3. (Optional) Preconfigure system installation information.	Use the `plrgconfig` file or the name service to preconfigure installation information, for example, the `locale`. Then, you are not prompted to supply that information during the installation.	"To Preconfigure System Installation Information" on page 26
4. Prepare the system for a custom QuickStart installation.	Create a QuickStart directory, which involves doing the following: ■ Adding rules to the `rules` file ■ Creating a profile for every rule ■ Testing the profiles ■ Validating the `rules` file	Chapter 6, "Preparing Custom QuickStart Installations"

EXAMPLE 7–2 Task Map (Continued)

Task	Description	Instructions
5. *For network installations only.* Set up the system to install the software over the network.	To install from a remote CD image, set up the system to boot. Install the software from an installation server or a boot server.	Chapter 9, "Preparing to Install PlirgSoft Software Over the Network"

For additional task map examples, see "Organize by Task Map" on page 118.

Using Flow Diagrams to Organize Tasks

A *flow diagram* organizes related, optional, and conditional tasks in a graphical format. If a task is complex and requires a high-level overview, you can use a flow diagram to lead readers through the task. A flow diagram can also provide cross-references to instructions for completing each task or subtask. For more information about flow diagrams and to view an example, see "Organize by Flow Diagram" on page 116.

Use Continuous Prose for Some Tasks

If a task is not suitable for being explained in a procedure, use continuous prose to write the instructions. The following two examples show task-oriented material that is written as continuous prose.

EXAMPLE 7–3 Task Written as Continuous Prose

8.4.1 Verifying the Result of an Asynchronous Operation

To enable your application to perform different actions depending on whether an asynchronous operation succeeds, verify the result of the operation. For example, verify the result of an asynchronous operation to notify the user or to perform another recovery action if the operation fails.

To verify the result of an asynchronous operation, call the `get_except` function of the `Waiter` class.

The `get_except` function returns one of the following values:

- If the operation failed, `get_except` returns a pointer to an instance of the `ExceptionType` class. This instance provides information on why the operation failed.
- If the operation succeeded, `get_except` returns `NULL`.

EXAMPLE 7–4 Another Task Written as Continuous Prose

12.3.1 Activating Access Control for the PlirgPak Manager Platform

Access control is set active or inactive during installation. If you want to enforce access control in your application, make sure that access control is active.

To determine whether access control is active, call the `get_access_control_switch` function of the `ACAccessControlRules` class.

To activate access control, call the `set_access_control_switch` function of the `ACAccessControlRules` class. In the call to `set_access_control_switch`, specify the access control switch status as `emAccessControlOn`.

To deactivate access control, call the `set_access_control_switch` function of the `ACAccessControlRules` class, specifying the access control switch status as `emAccessControlOff`.

Do Not Use Procedures for Command Explanations

Although commands do perform actions, merely using a command and its options does not typically constitute a complete task that the user performs. Therefore, in most cases you should not use procedures for the sole purpose of documenting command options. This section describes strategies for presenting commands and their options.

When Not to Describe Commands in a Procedure

Do not use procedures or steps solely to explain a command's syntax and options. Instead of a procedure, use an alternative means such as a man page or a syntax explanation in a reference section. Include index entries and descriptive section headings to point readers to the command's functionality.

Use an explanation to present a command in the following situations:

- If the information presents a command with many options and the information is not focused on a particular action. For example:

Incorrect:

How to List Resources

- Use the `list` command to list resources.

  The `list` command enables you to list resources. The syntax of the command is as follows:

  `list -`*option*

`-c`	Lists all virtual server classes.
`-l`	Lists all listen sockets.
`-r` *resource*	Lists the specified resources.
`-v`	Lists all virtual servers.

Correct:

`list` Command

The `list` command enables you to list resources. The syntax of the command is as follows:

`list -`*option*

`-c`	Lists all virtual server classes.
`-l`	Lists all listen sockets.
`-r` *resource*	Lists the specified resources.
`-v`	Lists all virtual servers.

- If the user will always perform a command as part of a larger task.

  For example, some commands are usually used to get information that is necessary to complete a larger task rather than by themselves as a single action.

When to Describe Commands in a Procedure

Use a procedure to present a command in the following situations:

- If the user is likely to perform a command as a single action and not always as part of a larger task. The action does not necessarily have to be a single step.

For example:

How to Check the Status of a Printer

1. **Log in to any system on the network.**
2. **Check the status of the printer.**

   Only the most commonly used options of the `lpstat` command are shown here. For other options, see Chapter 4 in the *PlirgSoft Reference Manual.*

`-d`	Shows the system's default printer.
`-D` *printer-name*	Shows the description of the specified *printer-name*.
`-t`	Shows status information about the LP print service, including the status of all printers, such as whether they are active and whether they are accepting print requests.

- If the information about the command is contained in a task-based section or chapter.

  For example, consider a system administration guide with a chapter that contains 3 pages of overview information and 10 tasks. Eight of the tasks are procedures that consist of multiple steps and require the execution of at least two commands. The two tasks that solely require issuing a command should also be presented as procedures.

- If you are emphasizing the subset of the command's options that is required to perform the operation that you are describing.

  Incorrect:

  ### Manage Resources, Resource Types, and Resource Groups

  - Use the `scrgadm(1M)` command.

    ```
    scrgadm -a | -c|-r -j {resource | -g resource-group | -t resource-type} \
    [-h node-list] [-x extension-property=value, ...] [-y standard-property=value, ...]
    ```

    [Explanation of *all* options and option arguments in the command.]

  Correct:

  ### How to Create a Resource

  1. **Create a resource group to contain the resource that you are creating.**

     ```
     # scrgadm -a -g resource-group [-h node-list]
     ```

     [Explanation of the replaceable values in the step]

  2. **Add the resource to the resource group that you created in Step 1.**

     ```
     # scrgadm -a -j resource -t resource-type -g resource-group \
     [-x extension-property=value, ...] \
     [-y standard-property=value, ...]
     ```

     [Explanation of the replaceable values in the step.]

     For more information, see the Chapter 4 in the *PlirgSoft Reference Manual.*

Writing Procedures

A procedure is usually an ordered set of steps. However, a procedure can include only one step. A procedure can include prerequisites. A procedure can also be preceded by introductory text or by cross-references to overview or supplementary information.

A procedure can also be followed by one or more examples and pointers to the next procedure or next topic that needs to be addressed.

To write effective procedures, follow these guidelines:

- Write procedures that are easy to follow.
- Place procedures appropriately.
- Use procedure headings appropriately.
- Use one method to describe a single procedure.

Write Procedures That Are Easy to Follow

To help readers understand and follow procedural content, use these guidelines when writing procedures:

- Make sure that you establish the entire context in which the procedure is done.

  You cannot assume that readers have read surrounding paragraphs or procedures preceding the current procedure. Do not assume that readers have already opened the screen or the file that previous sections or procedures discuss.

- Try to write no more than 10 steps for each procedure.

  If a procedure is a long, single series of steps, the procedure might be too complex. If a procedure is too long, review the user task analysis to see whether you can divide the task into two or more smaller procedures.

  Do not break up a long procedure if you cannot logically divide the steps. Therefore, do not separate a single procedure into smaller, less comprehensible procedures just to meet the recommended number of steps.

- Do not number single-step procedures.
- Include any prerequisite information.

  See "Including Prerequisites" on page 141 for details.

- Provide explanatory text and visual cues.

  Tell readers what is to happen after each step. For example, if a window opens as a result of a step, state that the window opens, and refer to the window by its name.

- Include all required information.

  Readers do not like to search through pages or go to another document to find required information. In such cases, duplicate the required information.

  However, if procedural content is common to many procedures in a task, include the procedure at the start of the task. Then, in subsequent procedures, cross-reference the common procedure.

- Do not provide a detailed description of each window, menu, or field in an interface.

  Most users explore an interface with a specific task in mind. Therefore, in procedures, only describe the parts of the interface that are necessary to complete the task.

  In addition, given their task orientation, users typically do not read or need separate sections that list and describe each part of an interface.

- Do not use illustrations in place of procedural information.

  An illustration providing an overview of the areas in a window can help orient readers. However, do not provide illustrations of the interface or portions of it in place of step-by-step procedures. In addition, do not provide procedural information in the callouts of an illustration.

- Do not repeat overview information or information that is not related to the task.

  Place overview information in a section or paragraphs before the procedure. Cross-reference any related, detailed supplementary information that supports the procedure.

- After the procedure, add one or more examples if doing so can help readers.

  For information about including examples, see "Providing Examples" on page 142. Then, consider pointing readers to the next procedure or next topic to be addressed.

Place Procedures Appropriately

To place procedures appropriately and consistently in a document structure, follow these guidelines:

- Put one or more procedures inside a section.

  Do not construct a chapter that contains only procedures and no sections.

  Place any introductory text or overview information that relates to one or more procedures under a section heading. The hierarchy for procedures within a section is as follows:

  1. A section heading
  2. Optional introductory text
  3. The procedure
  4. The steps
  5. Another procedure
  6. The steps

Generally, steps follow a procedure heading. If the procedure requires context, insert the text between the procedure heading and the first step. Introductory text is optional and should not merely repeat the text of the procedure heading. For example, if the procedure is titled "To Delete an Alarm," do not include as the sole introductory text "This procedure describes how to delete an alarm."

If introductory text is appropriate, use one of these constructions:

- Full paragraph
- Complete sentence that ends with a period
- Complete sentence that ends with a colon

- Try to put procedures under a first-level section or a second-level section, not a third-level section.

  Place related procedures at the same level.

  Procedures usually belong higher than a third-level section in a document's hierarchy. For third-level procedure headings to appear in the table of contents in print, you might need to contact your production group.

- Do not nest a procedure within another procedure.

  For example, do not put second-level procedure headings under first-level procedure headings. Put second-level procedure headings under second-level section headings.

Use Procedure Headings Appropriately

Procedure headings identify single-step procedures and ordered sets of steps. Any text that includes at least one step is a procedure. When writing procedure headings, follow these guidelines:

- Identify software procedures with procedure headings that start with "To" or "How to."

  To Customize User Files
  How to Customize User Files

  If you know which infinitive phrase your readers are accustomed to seeing, use that phrase. Be sure to use the phrase consistently within the document or documentation set.

Note – Some documentation styles number each heading and begin each heading with a gerund. For example: "3.3 Initializing POST."

- Write succinct yet meaningful procedure headings.

  See "Headings" on page 58 for instructions on writing effective headings.

- Do not place a colon at the end of procedure headings.

- If you provide instructions for more than one way to perform a procedure, indicate the method (in parentheses) in procedure headings, using this wording:
  - To Install the System (Command Line)
  - To Install the System (GUI)
  - To Install the System (Browser Interface)

Use One Method to Describe a Single Procedure

If readers can perform a procedure in more than one way, show only one method in a single procedure. For example, do not mix steps that use a command-line interface with steps that use a graphical user interface (GUI) in the same procedure.

Choose one method of presentation, command-line interface or GUI, that best suits the needs of your readers and the organization of the document.

An alternative is to present each method separately. Again, consider the needs of your readers and the organization of the document. Some of the more common ways to present each method separately are as follows:

- Put command-line procedures in one chapter and GUI procedures in another chapter.

  Ensure that each chapter title identifies the specific method.
- Put related command-line procedures in one section and related GUI procedures in another section.

  Ensure that each procedure heading identifies the specific method.

  To Start a Compilation (GUI)
  To Start a Compilation (Command Line)
- Primarily use the GUI method, showing pertinent screen captures of GUI windows within the procedure.
  - Use screen captures as a supplement to steps, not as a substitute for steps.

    Do not overuse screen captures. Provide visual cues only as necessary, and explain what happens after each step. See "Creating Quality Screen Captures" on page 243 and "Protecting Confidential Information in Examples" on page 200 for additional information about using screen captures.
  - If the entire procedure could be performed using the command line, include a command-line example that shows the same actions at the end of the procedure.
  - If only one step has a command associated with it, consider adding a Note or Tip following the step that shows the command.

1. **Click in the text where you want the symbol to appear.**

2. **Choose Symbol from the Items menu.**

   The Insert Symbol window appears.

   Tip – Alternatively, you can display this window by typing `show symbols` at the command line.

   ---

3. **Double-click the symbol name.**

   The symbol is inserted.

4. **When you have finished inserting symbols, close the Insert Symbol window by pressing Alt-R.**

If you provide both GUI and command-line information, provide readers with enough information to choose one method over the other method. For example, explain that GUI procedures offer a simple interface and instant verification. Explain that command-line procedures are often preferred by users who are familiar with using code and want automation in a script.

For guidelines on writing command-line procedures and GUI procedures, see "Write Meaningful Steps" on page 158. For additional guidelines on writing GUI procedures, see Chapter 14, "Documenting Graphical User Interfaces."

You can use illustrations to show how to perform hardware procedures. See Chapter 12, "Working With Illustrations," for illustration guidelines.

Writing Steps

When writing steps, determine what a user needs to do first, next, and last. Write clearly so that a user understands exactly what to do. To write concise steps, follow these guidelines:

- Order and number the steps.
- Make each step short and equivalent to one action.
- Write each step as a complete sentence in the imperative mood.
- Write meaningful steps.
- Use branching of steps appropriately.

Order and Number the Steps

When ordering and numbering steps, follow these guidelines:

- Present information in a logical order.

  For example, make sure that a step that requires certain information appears after the instruction that explains how the reader acquires that information.

  Incorrect:

  2. **In the *sge-root*/`default/common/act_qmaster` file, replace the current host name with the new master host's name.**

     This name should be the same as the name returned by the `gethostname` utility. To get that name, type the following command:

     ```
     # sge-root/utilbin/$ARCH/gethostname
     ```

  Correct:

  2. **Determine the new master host's name.**

     ```
     # sge-root/utilbin/$ARCH/gethostname
     ```

  3. **In the *sge-root*/`default/common/act_qmaster` file, replace the current host name with the new master host's name.**

- Do not use a number if a procedure has only one step.

  If you assign a numeral "1" to a single-step procedure, readers might think that one or more steps are missing. Consider providing a glyph in place of a numeral.

- If procedures include two or more steps, use numerals to number steps.

  Use letters for sequential substeps.

- Indicate optional steps by including the word "Optional" in parentheses.

  3. **(Optional) Reboot the system.**

  Do not identify a step as optional if any user needs to complete the step for the procedure to be successful.

- If users must perform different actions based on the outcome of a step, use a bulleted list to show the alternatives.

  Use letters for substeps. Also note the branching used in this example.

  1. **Determine whether you need to remove the disk drive.**
     - If no, go to Step 2.

- If yes:

  a. Gently pull back on foam piece 2.
  b. Slide the disk drive out of foam piece 2.
  c. Place the drive on the antistatic mat.

2. **Carefully lift the component from the unit.**

Make Each Step Short and Equivalent to One Action

A user can more easily follow a procedure when each step is short and explains one action. To help a user understand what to do in each step, follow these guidelines:

- Try to use no more than 20 words to write each step.

  Hyphenated terms and multiword product names count as one word. Explanatory text under the step text is not part of this word count.

  Incorrect:

  6. **Type the email address of the recipient in the To field of the Compose window, using spaces or commas to separate multiple addresses.**

  Correct:

  6. **Type the recipient's email address in the To field of the Compose window.**

     Use spaces or commas to separate multiple addresses.

- Place any explanatory text in a separate paragraph under the step text, and keep the explanatory text as short as possible.

  Incorrect:

  1. **Choose File → Find.**
  2. **After you make this selection, the Find dialog box appears.**
  3. **In the Find field, type the text that you want to locate. Because this search is case-sensitive, be sure to use appropriate capitalization.**

  Correct:

  1. **Choose File → Find.**

     The Find dialog box appears.

  2. **In the Find field, type the text that you want to locate.**

     This search is case-sensitive.

- Do not bury steps in a paragraph.

  Incorrect:

  Become superuser, stop the NIS server, change your files, and restart the NIS server.

  Correct:

  1. **Become superuser.**

  2. **Stop the NIS server.**

     ```
     # /etc/init.d/yp stop
     ```

  3. **Make the necessary changes to your files.**

  4. **Restart the NIS server.**

     ```
     # /etc/init.d/yp start
     ```

- Write about only one action in each step.

  Exceptions to this guideline include the following:

  - You conclude a step with "and press Return" (or "and press Enter") because that keystroke is a necessary component of the step.

    However, if all steps in a procedure conclude with "and press Return," do not repeat this instruction at the end of each step. Instead, explain in introductory text to the procedure that users must press the Return key after each step. If the book contains many procedures, explain this use of the Return key in an introductory chapter or the preface.

  - You begin a step with a common instruction such as "Log in as," "Log in to," or "Become superuser," followed by another short instruction.

    However, this combination of steps depends on the audience and the subject matter. Novice users might need this type of instruction, while experienced users would not.

    For novice users, provide this level of detail:

    1. **Become superuser.**

       ```
       $ su
       Password:
       ```

    2. **Reboot the system.**

       ```
       # reboot
       ```

    For experienced users, provide general instructions:

    1. **Become superuser and reboot the system.**

- Make sure that you include steps for all actions that the user must perform.

  Incorrect:

  To Add a Menu for All Users

  To add a menu for all users, use the File Manager to create the folder where you want to add the menus.

  Correct:

  To Add a Menu for All Users

  1. **In a File Manager window, type in the Location field the location where you want to add the menu, then press Return.**

     For example, to add a menu to the Applications menu, you would type `applications-all-users:///`.

  2. **From the File menu, choose New Folder.**

     An untitled folder is added to the View pane. The name of the folder is selected.

  3. **Type a name for the folder, then press Return.**

     The information file for the location that you specified in Step 1 is automatically updated with the details of the new menu. The name of the folder is displayed as the name of the menu.

Write Each Step as a Complete Sentence in the Imperative Mood

Verbs do most of the work in instructions. Reserve participles and gerunds for lists. When writing steps, follow these guidelines:

- Write each step as a complete, correctly punctuated sentence.
- Phrase the step as an action rather than a question or statement.

  Incorrect:

  1. **Do you have an account on this system?**

     What you do next depends on whether you have an account on the system.

  Correct:

  1. **Type your password.**
     - If you have an account on this system, type your password.
     - If you do not have an account, contact your administrator.

- Ensure that each step contains an active verb in the imperative mood.

  Put the verb at the start of the step unless you are explaining why, how, or where an action takes place. You might clarify a step in order to do one of the following:

  - Qualify the verb.

    Gently lift the I/O board up and out of the unit.

  - Provide information to orient readers.

    In the Add Attachments window, click Add File.

  - State a condition.

    If the card's I/O address conflicts with another device, change the I/O address according to the manufacturer's instructions.

  - Show the desired outcome or reason for the action.

    To secure the board to the unit, tighten both screws.

  - Stress the importance or consequence of an action.

    To shut down the system, type `shutdown`.

- Do not use command names as verbs.

  Incorrect:

  4. `cd` **to the new directory.**

  Correct:

  4. To change to the new directory, type `cd` *directory-name*.

Write Meaningful Steps

To write complete steps that are effective, do the following:

- For GUI procedures, clearly state what users need to do to interact with the interface.

  State what data the user needs to type in a text field, which menu option the user must choose, which radio button option the user must select, and so on. For example:

  Incorrect:

  5. Send the message to the recipients.

  Correct:

  5. Click Send.

  The message is sent to the recipients.

  Note that in GUI procedures, the task ("Send the message...") is not the focus of the step, as is the case with command-line procedures.

- For command-line procedures, make the task, not the command, the focus of the step. Follow the step immediately with command syntax, if applicable.

  Incorrect:

  6. Type `ufsrestore` **and press Return.**

  Correct:

  6. Verify that you successfully backed up the system.

  ```
  # ufsrestore -t
  ```

- For command-line procedures, follow the step and the command line with a description of the command options and variables that directly relate to the step.

Note – If the command line contains only one or two self-explanatory variables and no command options, you can choose not to define the variables. For example, if *filename* and *username* are the only variables, you can choose not to define these variables.

To present command options and variables, consider using one of these formats:

- A sentence that explains one or two command options or variables

  4. To examine a crash dump, use the `mdb` **utility.**

  ```
  # /usr/bin/mdb -k  crashdump-file
  ```

  where *crashdump-file* is the name of the crash dump file for the operating system. The `-k` option specifies kernel debugging mode.

  The use of lowercase "where" is acceptable in this instance because this format is an industry standard.

- A two-column format that describes two or more command options or variables

  4. Shut down the system.

  ```
  # shutdown -i0 -gn -y
  ```

`-i0`	Shuts down the system to init state 0 (zero), the power-down state.
`-gn`	Notifies logged-in users that they have *n* seconds before the system begins to shut down.
`-y`	Specifies that the command runs without user intervention.

  Do not use "where" to introduce the options or variables that are being described using this format because lone words can cause translation problems.

- Explain to readers why they are to skip a step or jump to a step.

  3. **Determine whether you want the partition table to be the current table.**
     - If you want to change the displayed partition table, type `n` and go to Step 4.
     - If you want to use the current partition table, type `y` when prompted:

     ```
     Okay to make this the current partition table [yes] y
     ```

- When providing an instruction in a step, make sure that you provide all the information that your audience needs to complete the step.

  For example, if you instruct the user to stop a server or to edit a file, provide information about how to do so.

Use Branching of Steps Appropriately

Use branching if the action to take at a particular step in a procedure differs depending on the user's situation or desired outcome. Follow these guidelines to determine whether a step requires the use of branching:

- Use branching if the procedure is the same for many cases and differs only at one or two steps.

  2. **Format the diskette.**
     - To format the diskette for a UFS file system, type `fdformat` and press Return.
     - To format the diskette for a Windows file system, type `fdformat -w` and press Enter.

- Indicate the branching condition in the main step text, not by using the word "or."

  Incorrect:

  8. **Restart NFS.**

     ```
     # /etc/init.d/nfs/restart
     ```

     OR

     ```
     # /etc/rc3.d/s60nfs restart
     ```

  Correct:

  8. **Restart NFS by typing one of the following commands:**
     - `/etc/init.d/nfs/restart`
     - `/etc/rc3.d/s60nfs restart`

- If several steps contain branches, consider splitting the procedure into two separate procedures.

- If the branching condition applies to the whole procedure, use two different procedures.

  For example, if the procedure has several steps that provide alternatives for HTTP and FTP protocols, create two procedures. Write one procedure that shows how to accomplish the task through HTTP and the other procedure through FTP.

- If a particular condition requires a substitution in most of the steps in the procedure, provide that information in a Note.

  For example, tell readers to use the default directory if their default directory is different from the directory provided in the steps.

  Put the Note at the beginning of the procedure, not in the branches of the steps.

- If a step is optional, do not use branching.

  Incorrect:

  6. Determine whether you want to make this printer the default printer.
  - If no, go to Step 7.
  - If yes, select Default.

  Correct:

  6. (Optional) To make this printer the default printer, select Default.

  If one branch states to proceed to the next step, you might be able to use an optional step, not a branch. However, do not use an optional step if any user needs to complete the step for the procedure to be successful.

- Do not use branching to provide all of the possible actions in a confirmation step at the end of a procedure.

  For example, many graphical user interfaces provide OK, Apply, and Cancel buttons in each dialog box. If you have described the actions of these buttons in a central location, do not repeat this information at the end of every procedure. Assume that the user wants to confirm the settings in the dialog box and just say "Click OK."

- If a user must know certain information to determine which branch to follow, include the process by which the user can find out that information.

  Incorrect:

  3. Ensure that the `.html` file is complete.
  - If the file is complete, post the file on the internal web site.
  - If the file is not complete, see Appendix C.

  Correct:

  3. Use a text editor to ensure that the `.html` file is complete.
  - If the file is complete, post the file on the internal web site.
  - If the file is not complete, see Appendix C.

Checking for Structural Problems

Use the initial user task analysis to guide your procedure writing. As you write, you might discover that you have to further break down or combine some procedures. This section describes some signs of a possible need for restructuring.

Duplicate Series of Steps

If two or more procedures begin with the same series of steps, consider creating a separate procedure with the shared steps and then cross-referencing to that procedure in the related procedures.

For example, suppose you have several procedures that are accomplished through a web page deep in the application's hierarchy. To describe how to get to the page requires four steps. You can create a separate procedure such as "To Access the Modify Objects Page." Then, at the beginning of each modification task, Step 1 can say, "If you are not already on the Modify Objects page, see 'To Access the Modify Objects Page' on page 8."

Nearly Identical Procedures

Look for two or more procedures that are alike except for one or two steps that require a different value or choice. Consider combining the procedures into one procedure. Provide information in the steps that require alternative choices.

For example, a word processing application might use the same basic procedure to create generated lists such as tables of contents, lists of tables, and lists of figures.

Rather than having separate procedures for each type of generated list, you can provide one procedure that describes how to create generated lists. Then, in the relevant steps you can provide the specific file names related to the type of list the user is creating.

See "Use Branching of Steps Appropriately" on page 160 for details.

Many Nested Substeps

Multiple levels of substeps in several steps in a procedure probably indicate that the procedure should be divided into separate procedures.

Procedures With More Than 10 Steps

As mentioned in "Write Procedures That Are Easy to Follow" on page 149, long procedures are difficult to follow. Look for a logical place to divide the procedure.

You might want to describe the overall procedure, followed by a task map or numbered list that describes the related procedures.

Several Single-Step Procedures

The presence of many single-step procedures might indicate a few different structural problems:

- If many of the procedures describe the same basic action, you might be able to collapse the procedures.

  For example, suppose you have separate single-step procedures for opening different applications from a front panel. You might want to provide one single-step procedure with the heading "To Open an Application From the Front Panel." If necessary, provide a cross-reference from each application.
- If some of the procedures are related logically, you might be able to combine the procedures.

  For example, suppose you must add one line to a system file to set a printer resource and another line to set a scanner resource. You can provide a procedure with the heading "To Add Peripherals to the `.Nresource` File." The procedure contains the steps that are common to the procedures that are being combined. Then one step includes the different text lines for each peripheral.
- Examine the single-step procedures to ensure that all required steps are provided.

  For example, you might be assuming that the user is at a particular place in the GUI or has already logged in to the system.

Repeated Steps Indicated at the End of Repeated Actions

Instructions such as "Repeat steps 2 through 4" that are placed at the end of a set of steps indicate a structural problem. Instead, indicate the repetition at the start of the steps to be repeated. Sometimes, this strategy results in nested substeps, but it also shortens the overall number of steps and gives the user sufficient context in the form of an advanced warning.

Incorrect:

3. Determine which data links are configured on a host.

```
# dladm show-link
```

4. Identify which VID to associate with each data link on the system.

5. Create PPAs for each interface to be configured with a VLAN.

6. Repeat steps 3 through 5 on each system.

Correct:

3. On each system, perform the following steps:

   a. Determine which data links are configured on a host.

   ```
   # dladm show-link
   ```

   b. Identify which VID to associate with each data link on the system.

   c. Create PPAs for each interface to be configured with a VLAN.

◆ ◆ ◆ CHAPTER 8

Writing for an International Audience

More and more business transactions and communications occur over the World Wide Web, which is an international medium. Writing documentation that can easily be translated into other languages and delivered to audiences in other countries is becoming a mandate for the computer industry. Fortunately, the guidelines that you need to follow when writing for an international audience also apply to good technical writing in general. These guidelines can help you avoid producing documentation that is inadvertently confusing or offensive.

Internationalization involves creating a "generic" document that can be used in many cultures or easily translated into many languages. *Localization* involves converting a document that is specific to a particular language or culture into one that is specific to a different language or culture.

Working closely with translators and localization experts who are based in the countries to which you are exporting is important. See "Internationalization and Localization" on page 423 for books about developing software and preparing documentation for the international market. See "Internationalization and Globalization" on page 350 for a discussion of management issues related to the global market.

This chapter discusses the following topics:

Guidelines for Writing for Translation

Following the basic guidelines for good technical writing is important when you are creating documentation. Following these guidelines can also help you avoid producing documentation that is confusing or offensive to translation vendors or to readers from other cultures.

Follow these basic guidelines when you are preparing documentation for an international audience:

- Keep the documentation culturally neutral by avoiding elements that are difficult to translate or hard to read by people whose native language is not U.S. English.

  These elements are explained throughout this chapter. At many companies, documentation that is written in English is often distributed worldwide.

- Have your document edited, if possible, before giving it to a translation vendor.

  Many complaints from translation vendors concern basic errors in the English version of a document, such as typos and inconsistent term usage.

- Be aware that text expansion can occur when a document is translated.

  A document can increase in size by up to 25 percent, which can substantially expand the breadth of a hard-copy document. Such expansion can affect the binding, packaging, and shipping constraints. So, aim to be clear but concise in your writing.

Cultural and Geographic Sensitivity

More than ever, technical communicators need to think globally. Conventions that are standard in one country might be handled differently in other countries. Use the following guidelines when writing for an international audience.

Use Culturally Neutral Examples

When including examples in your text, keep the following guidelines in mind:

- Avoid using examples that are culturally bound, such as names of places, public figures, or holidays that might be unrecognizable to people living in different countries.

  If you do use examples that are culturally bound, use examples that represent a variety of cultures or that are internationally recognized. For example, you may use international cities, such as Paris, New York, Tokyo, London, and Hong Kong.

- Avoid political or religious references.
- Avoid gender-specific references.

Include International Date, Time, and Contact Information

When including dates, times, addresses, and telephone numbers, follow these guidelines:

- Be aware that dates are displayed differently in different countries.
  - **Month, day, year**. Used mainly in the United States.
  - **Day, month, year**. Used in Europe and Australia.
  - **Year, month, day**. Used in Asia. This format is also used by the International Organization for Standardization (ISO) standard for numeric representation of dates.

  For clarity, write out dates. For example, write "6/28/03" as "June 28, 2003." If abbreviations are necessary, define them and then use them consistently.
- Be aware that times are displayed differently in different countries.

  Time formats that use a 12-hour clock and the *ante meridiem, post meridiem* (a.m. and p.m.) system are not universally understood. Consider using a 24-hour system or describing the time in relation to the time of day. For example, you can write "1:00 p.m." as either "13:00" or "1:00 in the afternoon."
- Ensure that address or telephone information is complete.

  Include telephone country codes, area codes, and time zones when you provide phone numbers and calling hours in a document that might be distributed internationally. Use "+1-" before the area code for phone and fax numbers within the United States. Be aware that any toll-free telephone numbers that you provide might incur a charge or be unreachable from most countries outside the United States. Do not forget to specify the time zone if your readers are from more than one time zone.
- Be careful when including information about warranties or technical support.

  Be specific as to which countries honor the warranties or offer technical support, if possible.

Avoid Informal Language and Styles

The following guidelines provide some tips for using the proper tone when writing for an international audience.

- Avoid humor.

  What might be funny in one language, whether an illustration or written text, might be obscene in another language. Humor is strictly cultural, and it cannot be translated easily from one language to another language.
- Avoid irony.

  Even native speakers of English have difficulty discerning irony in writing.
- Avoid idioms and nonstandard colloquialisms.

  Nonnative English speakers often misunderstand idioms and nonstandard colloquialisms because they interpret these terms literally. For more information, including a list of terms to avoid and suggested alternatives, see "Idioms and Colloquialisms" on page 107.

- Avoid metaphors.

  Some metaphors are acceptable if they are explained in the text. However, some metaphors can easily be replaced. The following guidelines can help you decide whether a metaphor is acceptable:

  - If you can easily replace the metaphor with one or more words, rewrite accordingly.

    Incorrect: The drawback is that deleted calendar components will reappear in the rebuilt database.

    Correct: The disadvantage is that deleted calendar components will reappear in the rebuilt database.

    Incorrect: The only pitfall to avoid is the use of any token names in the grammar that are reserved or significant in the parser.

    Correct: The only problem to avoid is the use of any token names in the grammar that are reserved or significant in the parser.

    Other possible rewordings for "pitfall" are "hidden danger" and "error."

  - If the metaphor precisely and concisely describes a subject and the metaphor cannot easily be replaced, use the metaphor but explain it in the text.

    If you use a metaphor, provide additional contextual explanation so that the translator does not misunderstand and translate the text incorrectly. If the translator is unfamiliar with the metaphor, the translator might have to guess at the meanings of key terms.

    For example, if you describe a file system hierarchy as a "tree structure," or if you describe a "parent-child" relationship, include an illustration or example. That way the translator understands that you do not mean these terms to be interpreted literally.

Definitions and Word Choice

Follow these guidelines to avoid wording that makes translators and readers uncertain of your intended meaning.

Avoid Jargon and Slang

If a term is not listed in a standard dictionary or a technical source book, do not use it. If a term is specific to your company but is not defined in the text or in the glossary, do not use it. If translators cannot look up an unfamiliar term, they might have to guess at its meaning.

Do Not Use Foreign Terms

Even though some foreign terms are listed in U.S. English standard dictionaries, do not use them in your documents. Nonnative English speakers might not understand foreign terms such as "vis-à-vis," "via," and "vice versa." In addition, when documents are translated, no equivalent terms exist for foreign terms in target languages.

Use Terms Consistently

Consistent terminology usage can help increase comprehension by nonnative English speakers. Follow these guidelines:

- Avoid using terms that can have several different meanings.

  For example, the word "system" can refer to an operating system (OS), a combination of OS and hardware, a networking configuration, and so on. If you do use such a term, ensure that you define it and use it consistently in a document. Ensure that you also add it to the glossary.
- Use terms consistently throughout a document.

  Synonymous terms in a document can be troublesome for a translator. The words "show," "display," and "appear" might seem similar enough to use interchangeably. However, a translator might think you used the different words deliberately for different meanings, and a translator might interpret the text incorrectly.

  Instead of using synonymous terms in a document, pick one term and use it consistently throughout your document or documentation set.

  Other inconsistent usage with which translators might have trouble include the following:
  - Down, crash
  - Menu option, menu item
  - Connector, port, plug
  - Output, result
  - Some, several, many, few
  - Platform, architecture, system
  - Scroll list, scrolling list, scrollable list
  - Executables, executable program, executable code, executable application, executable file
- For literal command names or other computer terms, include a phrase such as "the ... command" or "the ... function" on first reference. Including the surrounding article and term can clarify all references.
- Use uppercase and lowercase letters consistently in like elements throughout a document.

  Using consistent case helps a translator determine the proper interpretation of a term. The use of consistent case is significant for reserved keywords, class names, and variables.
- Be careful about using the word "available" because it has multiple meanings.

  Where possible, use definitive words such as "active" or "valid."

  Incorrect: The role determines the user's available permissions.
  Correct: The role determines the user's valid permissions.

  Incorrect: After you add the entry, the Edit button becomes available.
  Correct: After you add the entry, the Edit button becomes active.

However, the following are acceptable uses of "available" and "unavailable":

Further information is available at `http://www.plirg.com`.

Go to First Page (Unavailable) (as alternative text for the disabled version of a vertical pagination icon in a table)

If additional disk space is unavailable, the program fails.

Use Abbreviations and Acronyms Judiciously

Many languages do not have abbreviations and cannot accommodate them. Use abbreviations and acronyms judiciously in your document, and do not arbitrarily create them. Follow these guidelines:

- Use industry-standard abbreviations and acronyms in your documents, but define them at first use in the text.

  If you use abbreviations and acronyms in your document, define them the first time you use them in text. Provide acronyms and abbreviations at the end of the book as part of the glossary, in an appendix, or in a separate list of abbreviations. When you define the term, give the spelled-out version first, followed by the acronym or abbreviation in parentheses.

- Do not arbitrarily create your own abbreviations and acronyms.

  Abbreviations and acronyms that engineers create for their purposes do not necessarily belong in your document. Nor should you create them without careful thought. "DT" for "directory tree" and "MS" for "Master Server" are examples of arbitrary abbreviations that should not be used. However, new technology often requires the creation of new terms and possibly new abbreviations and acronyms.

- Avoid using abbreviations and acronyms in the plural form.
- Do not use Latin abbreviations such as e.g., i.e., vs., op. cit., viz., and etc.

  Latin abbreviations can cause problems for translators as well as comprehension problems for native and nonnative English speakers. Because these abbreviations might be obscure for your audience, you can achieve greater clarity by not using them.

For more guidelines on using abbreviations and acronyms, see "Abbreviations and Acronyms" on page 35.

Avoid Contractions

Avoid using contractions that are difficult to translate. If you do use contractions in your documentation, follow the guidelines described in "Contractions" on page 29.

Grammar and Word Usage

If you adhere to English grammar guidelines and use terms correctly in your documentation, you can eliminate much of the ambiguity that slows the translation process.

Follow These Grammar Guidelines

To help ensure correct usage, follow these grammar guidelines:

- Ensure that spelling and word usage are correct.

  Use electronic spelling checkers and copy editors to ensure accuracy.

- Do not omit articles such as "the," "a," and "an."

  Incorrect: Place screwdriver in groove.
  Correct: Place the screwdriver in the groove.

- Include the word "that" when it is used to introduce a restrictive clause.

  See "Differentiate Between Restrictive Elements and Nonrestrictive Elements" on page 94.

  Incorrect: Verify your configuration matches what is shown in the example.
  Correct: Verify that your configuration matches what is shown in the example.

- Avoid passive clauses, such as "the program was activated."

  See "Use Active Voice and Passive Voice Appropriately" on page 91.

- Do not put a list in the middle of a sentence.

  See "List Introduction Guidelines" on page 63.

- Check for the correct placement of prepositional phrases.

  Incorrect: Remove the filler panel from the slot with the pliers.
  Correct: Use pliers to remove the filler panel from the slot.

Use Words Precisely

Follow these guidelines:

- Be careful about using the same term in multiple grammatical categories, such as verb, noun, and adjective.

  Incorrect: Plug the plug into the wall outlet.
  Correct: Connect the plug to the wall outlet.

  Using "plug" as both a verb and a noun is confusing to translators. Translators might have to use a different term in each case. Also, do not use several terms to refer to the same thing.

- Be precise about using the words "when" and "if."

  Use "when" for an inevitable event and "if" for a conditional event.

  "*When* the prompt is displayed" implies that the prompt will be displayed.

  "*If* the prompt is displayed" implies that the prompt might or might not be displayed.

- Avoid using the word "may" unless you mean permission, as in "You may use these conventions in your code."

  Use the words "might" or "can" in place of the word "may." The word "might" indicates a possibility, and "can" means the power or ability to do something. A translator who must translate the English word "may" in text often chooses whether to translate it as "can" or "might." The original writer is in a better position to know which word is more accurate and should therefore use the correct word.

- Avoid using the words "there" and "it" at the beginning of a sentence or independent clause if those words take the place of the subject of the sentence and are followed by a linking verb such as "is" or "are."

  This construction delays the subject of the sentence, which can confuse translators.

  Incorrect: There are only a few troubleshooting tickets left.
  Correct: Only a few troubleshooting tickets are left.

  Incorrect: It is a simple path.
  Correct: The path is simple.

  Incorrect: In command-line mode, there are two ways to delete a file.
  Correct: In command-line mode, you can delete a file in one of two ways.

- Avoid ambiguous phrases.

  For example, "first-come, first-served" is ambiguous. If possible, rewrite as "in the order received" or "in the order in which they are received."

Use Modifiers and Nouns Carefully

Follow these guidelines:

- Be careful with compound modifiers.

  Compound modifiers can be hard to understand and to properly translate. You might need to rewrite the sentence or hyphenate the phrase, for example, "real time-saver," "real-time operation." See "Hyphen" on page 47 for guidelines for hyphenating terms.

- Avoid using general modifiers that might be interpreted in several ways.

  The translation of a term with multiple meanings requires the ability to discern the appropriate equivalent in the target language, based on the context. However, most translators do not have as much technical knowledge as engineers, and they might translate questionable terms incorrectly.

For example, the following sentence presents a difficult translation:

> The PlirgSoft Desktop Interface is an advanced Motif-based desktop with an easy-to-use interface that provides a consistent look and feel across software platforms.

The phrase "advanced Motif-based desktop" raises questions for a translator such as is it "a desktop that is based on advanced Motif" or is it "an advanced desktop that is based on Motif"? The sentence could be rewritten as follows:

> The PlirgSoft Desktop Interface is an advanced desktop system that is based on Motif. This system provides an easy-to-use interface that is consistent across software platforms.

- Do not use noun clusters of more than three nouns.

  Try to clarify noun clusters by using prepositions such as "of" or "for."

  Incorrect:

  In certain situations, the certificate chain verification process is disabled.

  Correct:

  In certain situations, the verification process for the certificate chain is disabled.

  An exception is a noun string that results when you use a three-word product name as a modifier. A product name that you use as a modifier counts as one word. See "Write Meaningful Steps" on page 158.

- Repeat the modifier in noun phrases that are joined or are linked together.

  Incorrect: You must set up a mail service on a new network or subnet.
  Correct: You must set up a mail service on a new network or new subnet.

- Repeat the main noun in conjoined noun phrases.

  Incorrect: You can access a new or existing network.
  Correct: You can access a new network or an existing network.

Limit the Use of Pronouns

The imprecise use of pronouns can cause confusion. Follow these guidelines:

- Avoid vague and uncertain references between a pronoun and its antecedent.

  A pronoun that forces a reader to search for an antecedent can frustrate or mislead the reader, as well as a translator. Ensure that the noun to which the pronoun refers is clear.

- Do not use the following words as pronouns:

All	Each	Many	One	Some
Another	Either	Neither	Other	Several
Any	Few	None	Own	

When these words are used as pronouns, their antecedent is unclear.

Incorrect:

These macros classify character-coded integer values. Each is a predicate that returns nonzero for true, 0 for false.

Correct:

These macros classify character-coded integer values. Each macro is a predicate that returns nonzero for true, 0 for false.

Incorrect:

Custom layout managers. To provide custom behavior that ensures the best GUI performance, write your own.

Correct:

Custom layout managers. To provide custom behavior that ensures the best GUI performance, write your own custom layout managers.

- Avoid using pronouns, such as "it," "its," "this," "they," "theirs," "that," "these," and "those," especially at the beginning of a sentence.

  Use these pronouns only when the noun to which the pronoun refers is clear.

  Unclear: This provides the following benefits.
  Clear: This support provides the following benefits.

Simplify Sentences

Break up sentences that contain more than two uses of the words "or" or "and."

Incorrect:

The software consists of four daemon processes that coordinate the scheduling, dispatch, and execution of batch jobs and monitor job and machine status, report on the system, and manage communication among the components.

Correct:

The software consists of four daemon processes. These processes perform the following functions:

- Schedule, dispatch, and execute batch jobs
- Monitor job and machine status
- Report system status
- Manage communication among the components

Numbers, Symbols, and Punctuation

Follow these guidelines to minimize confusion about the numbers and symbols in your documentation.

Clarify Measurements and Denominations

Most of the world uses the metric system, although many people in the United States are familiar only with the U.S. equivalents for the metric system. Also, number and currency formats vary worldwide. In many countries, commas and decimal points are used differently. As a courtesy to readers who use different numeric systems, follow these guidelines:

- When providing U.S. measurements, include the metric equivalent in parentheses if it is appropriate for the product you are describing.

  Most standard U.S. English dictionaries contain a U.S.-to-metric conversion chart under the "metric" entry.
- If you use the word "billion" or "trillion," explain the word in a footnote so that the exact value is clear to readers in all countries.
- If you are specifying prices, indicate the currency used.

  For example, write "USD" for United States dollars and "EUR" for the European euro.

Avoid Certain Symbols and Punctuation Marks

Different countries often have different conventions for symbols and punctuation marks. Follow these guidelines:

- Do not use these symbols:
  - The # symbol to indicate "pound" or "number"
  - A single quotation mark (') to indicate "foot"
  - Double quotation marks (") to indicate "inch"

  These symbols are not recognized in many countries outside the United States.
- Avoid using symbols such as "/" and "&" in text.

  Many symbols have multiple meanings, and translators might have difficulty deciding which meaning you intended. For example, the "/" symbol can mean "and," "or," "and/or," "with," "divide by," "root," or "path-name divider."

  See "Slash" on page 55 for acceptable uses of the "/" symbol.

Illustrations and Screen Captures

Follow these pointers to maximize the international appeal and comprehension of illustrations and screen captures.

Note – For legal reasons, be careful when providing examples in illustrations and screen captures. See "Protecting Confidential Information in Examples" on page 200.

Choose Illustrations to Communicate Internationally

When including illustrations, follow these guidelines:

- Use illustrations instead of text whenever possible to convey a complex concept.

  Ensure that the accompanying text complements the message conveyed by the illustration.
- Do not insert an illustration into the middle of a sentence.

  See "Placement in Relation to Sentences" on page 235.
- Remember that not everyone reads from left to right. If necessary, indicate the intended sequence that you want a reader to follow in the illustration.

EXAMPLE 8–1 Intended Reading Sequence in an Illustration

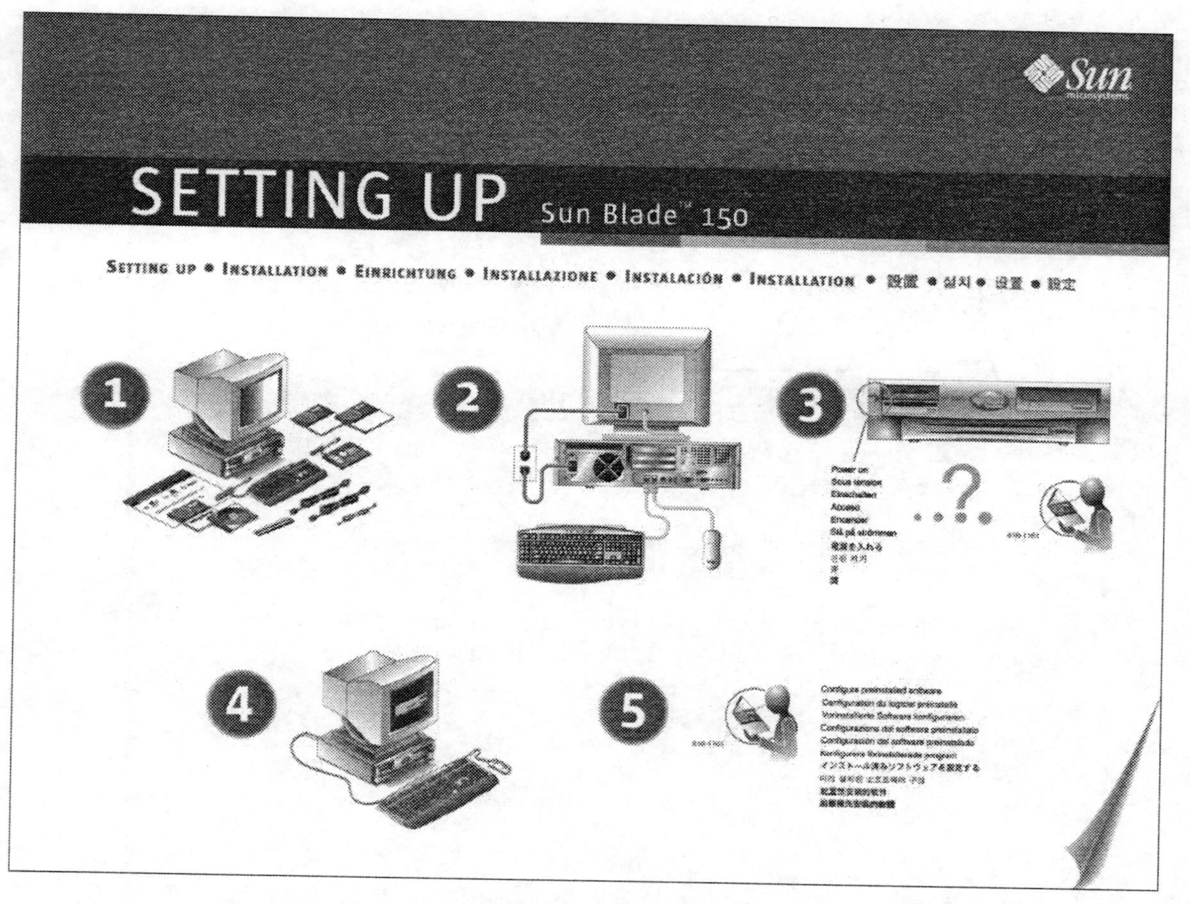

Use illustrations that are internationally acceptable. For example, almost every country has its own type of power connector. Instead of illustrating each type of connector, use generic connectors and receptacles, as in the following illustration.

EXAMPLE 8–2 Illustration of Internationally Generic Connectors and Receptacles

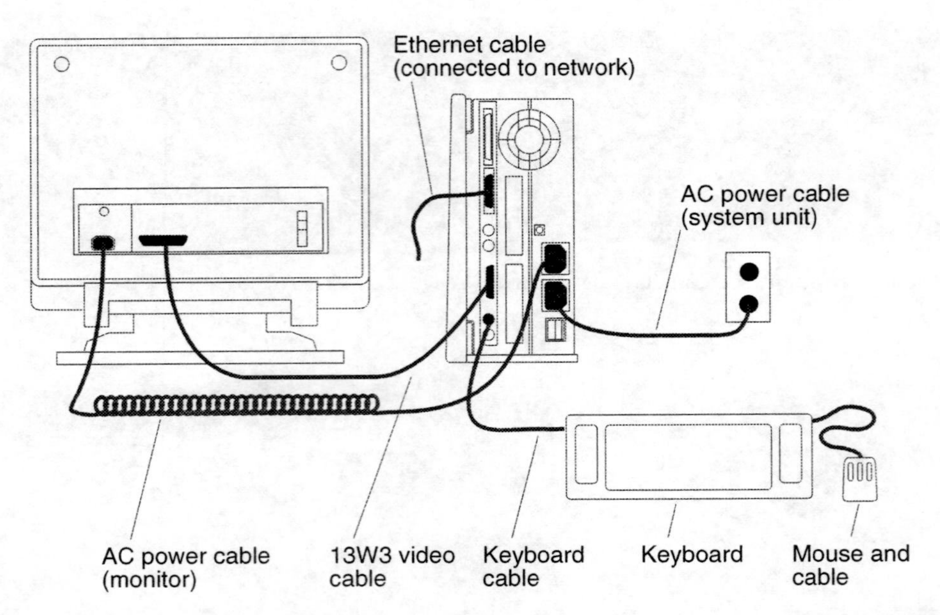

However, when describing various types of connectors and receptacles, illustrate and label the specific type used in each country, as in the following illustration.

EXAMPLE 8–3 Illustration of Internationally Specific Connectors and Receptacles

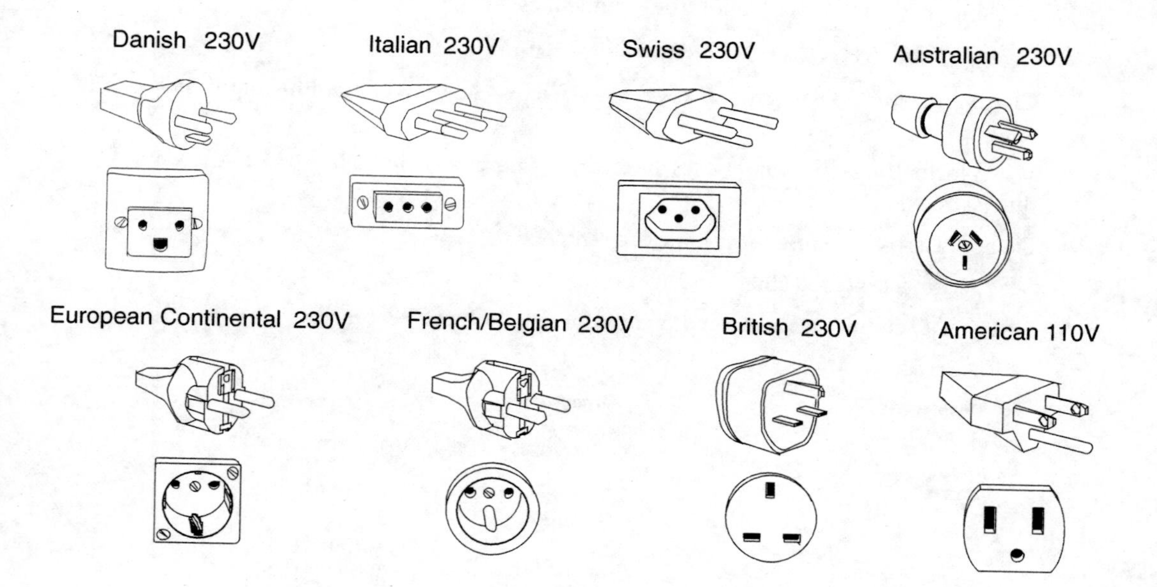

Create Callouts That Are Easy to Translate

A *callout* is a text element that defines a portion of an illustration or provides the reader with additional information about an illustration. When you create callouts in documentation that will be localized, follow these guidelines:

- Keep callouts short.

  Leave ample space, both vertically and horizontally, for callout text in illustrations.

  Translated text might require as much as 25 percent more space than English text.

- Make certain that the callouts correlate with the paragraph text.

  Use callouts instead of text if the concept can best be understood graphically and needs little explanation.

- Format callouts so that you can edit them separately from the illustration.

  For an example of an illustration with callouts that are editable (and scalable), see Example 12–7.

For more guidelines on creating callouts, see Chapter 12, "Working With Illustrations."

Use Charts and Tables

When using charts and tables, follow these guidelines:

- Use charts and tables to clarify essential information.

  Charts and tables are internationally recognized by readers as containing important material.
- When formatting charts and tables, be aware that text can expand by up to 25 percent during translation.
- Do not insert a table into the middle of a sentence.

  See "Writing Text for Tables" on page 70.
- Do not use sentence fragments with verbs for table column headings.

  Incorrect:

If the user asks...	Then you must respond by...
How do I do this?	Showing the user how to perform the task.

  Correct:

Questions	Answers
How do I do this?	Show the user how to perform the task.

- Do not use "Yes" and "No" or "True" and "False" in table data because these terms are problematic when translated into some languages.

  Instead of using "Yes" and "No" or "True" and "False" in table data, describe the cell contents based on the content in the table column heading. If describing the cell contents causes a problem with table column widths, consider using abbreviations or symbols and providing a key as a footnote to the table. That is, in the examples that follow, you could use "X" or "+" for "Applicable" or "Installed," and "N/A" or "-" for "Not applicable" or "Not installed."

Incorrect:
Conditions Applicable?
Device 1: Yes
Device 2: No
Device 3: Yes

Correct:
Conditions Applicable?
Device 1: Applicable
Device 2: Not applicable
Device 3: Applicable

Incorrect:
Components Installed?
Component 1: True
Component 2: False
Component 3: True

Correct:
Components Installed?
Component 1: Installed
Component 2: Not installed
Component 3: Installed

Use International Illustrations, Symbols, and Examples

Follow these guidelines to avoid confusion when using illustrations, symbols, and examples:

- Do not use a hand in a symbolic gesture.

  Almost any way that you position a hand can be considered an offensive gesture, depending on the culture, as shown in the following illustration.

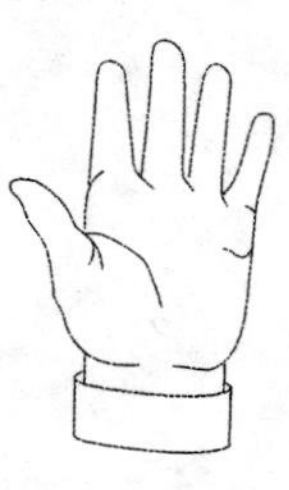

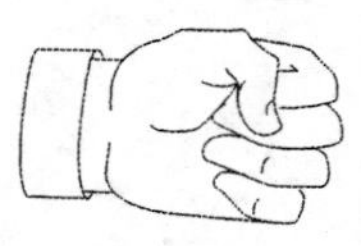

- When using code examples or showing screen captures, make sure that the machine names, login names, and system names are not culturally offensive.

  Instead of:

```
heavenly%
```

  Use a culturally neutral example:

```
system1%
```

- Avoid using road signs in illustrations because they differ from country to country.
- Avoid illustrations that relate to alcohol or alcohol-related material, because those types of illustrations might be offensive.
- Be careful when using everyday objects in examples.

  Be certain that the object exists in most countries. Be aware that the object might be interpreted in various ways in different parts of the world.

  For example, you can use a light bulb to indicate light, but not the concept of an inspiration.
- Avoid using trendy objects, historical references, or film, cartoon, or video characters.
- Avoid using animals, because they might carry symbolic significance.

  For example, some web pages use a dog to indicate a "fetch" function. While this connection might be clear to readers from Western cultures, it might not be understood by readers from other cultures. Dogs are not kept as pets or used for hunting and retrieving in all cultures.

- Do not use color to convey meaning.

  Color communicates different meanings in different cultures. For example, the color green is sometimes used to indicate paper money in the United States. However, the color green does not work in many other countries where money is not green. The color red can signify “stop” or “danger” in the United States. But in China, the color red can indicate happiness, and in France, red can denote aristocracy.

Legal Guidelines

Technical publications professionals need to follow legal guidelines that cover the proper usage and marking of trademarks and the protection of intellectual property. Intellectual properties such as trademarks, copyrighted works, and trade secrets are among a company's most valuable assets. Everyone who is involved in the preparation of materials that use trademarks or who creates materials subject to copyright has a responsibility for securing and protecting the copyrights and trademarks.

While this chapter mainly addresses legal requirements for print documents, legal guidelines for web pages, product documents on the web, and online help are also provided.

Note – In this chapter, the term "trademark" refers to a company's master brand and family brands, including associated logos, marks, and other designations. These brands, logos, marks, and other designations are used to identify a company's products and services.

When you see the term "registered trademark" or "service mark," information is specific to that type of trademark.

This chapter sometimes advises you to check with the legal department in your company. If you do not have a legal department, check with counsel specializing in trademark and copyright law. The International Trademark Association's web site (`http://www.inta.org`) provides information in this area, plus links to many U.S. and international web sites relevant to these issues.

This chapter explains how to use and designate corporate trademarks. It also explains how to protect your company's intellectual property and trade secrets. The topics in this chapter include:

Understanding Copyrights

This section provides the following information about copyrights:

- General copyright information
- Works that must be copyrighted
- Copyright notice
- Third-party copyrighted information

If you have copyright questions that are not addressed in this section, you might want to consult the web site of the United States Copyright Office at `http://www.copyright.gov`.

General Copyright Information

Copyright is a statutory right granted by federal law to protect original works of authorship, including literary, musical, dramatic, pictorial, computer program, and other types of intellectual works. U.S. copyright laws establish a single system of protection for all published and unpublished "original works of authorship fixed in any tangible medium of expression from which they can be perceived, reproduced, or otherwise communicated." With limited exceptions, no one may copy or reproduce, display, prepare derivative works, or distribute copies of copyrighted works to the public by sale, rental, lease, lending, or other transfer of ownership without permission of the copyright owner.

While no universal international copyright protection exists, United States copyrighted works might be recognized and protected in certain foreign countries under various international treaties and conventions.

This section discusses copyright duration, copyright registration, and trade secrets.

Copyright Duration

A copyright, unlike a trademark or patent, exists from the time of the creation of a work. For copyright purposes, a work is *created* when the work is fixed in a tangible medium of expression. For example, a chapter that you write becomes protected by copyright law the moment that you save the chapter to a file.

Even though a copyright notice is not required for a copyright to exist, certain legal benefits accrue when the notice appears on the work. A copyright notice shows ownership of your company's underlying copyright in the work to which it is affixed. Once affixed, a copyright notice cannot be lawfully removed by anyone other than the copyright owner. The presence of a conspicuous copyright notice is your company's first defense against any claim of copyright infringement.

A copyright exists for a finite number of years, generally the lifetime of the author plus 70 years. If the work is done as a work for hire for an employer, the copyright exists for 95 years from the date of publication. After that time, the work becomes part of the public domain.

Copyright Registration

Registering a copyright is not necessary for copyright ownership, but registration takes full advantage of the legal protection afforded copyrighted works.

To register a work, you must submit a copyright registration application to the United States Copyright Office. A copy of the work should be included as a deposit to be held on file with the Library of Congress. In the case of software, a portion of the source code is included. The material usually consists of the first and last 25 pages, with portions of the material blocked out as necessary to protect trade secret information. The names and the citizenship of all contractors (non-employees) who worked on the product or document are also required.

The following acts can usually occur only if a copyrighted work is registered:

- Filing a legally enforceable copyright infringement action in a court of law
- Obtaining statutory damages

  Statutory damages can be substantial and can be obtained only if the copyright infringement takes place *after* registration is secured.
- Obtaining attorney's fees
- Enlisting the United States Customs Service to bar the importation of illegal copies of the registered work

A federally registered copyright also gives the registrant many and various procedural advantages should the registrant want to take action against an infringer.

Copyright Compared With a Trade Secret

Copyright does not protect the ideas and concepts contained in a work, but only the expression of such ideas and concepts. Thus, copyright is not always the best single means of protection for material that contains valuable information that can be exploited.

Confidential business information and information that pertains to potential patent opportunities or applications are usually best protected as *trade secrets*.

To safeguard a trade secret, appropriately classify the information and use the corresponding confidential label. Restrict disclosure to authorized third parties. For more information, see "Protecting Confidential Information" on page 196.

What Should You Copyright?

The following works that are intended for distribution outside your company could benefit from a copyright notice:

- Publications such as books, articles, white papers, and brochures
- Advertising copy and news releases
- Photographs
- Catalogs
- Product labels
- Product documentation
- Web pages, product documents on the web, online help
- Software applications
- Source code and binary code for software products
- Source documents for which the source is distributed or published

You typically include a copyright notice with any documents or artwork intended for public distribution. If you do not know whether a copyright notice is appropriate for a particular type of work, check with your legal counsel.

Copyright Notice

A work that is suitable for copyright protection should contain a proper copyright notice.

A copyright notice consists of the following elements:

- The word "Copyright", the © symbol, or both elements

  If space is very tight and the text is not being converted to ASCII, use the © symbol.

  Do *not* use (c) or (C) to mean copyright.
- The year of first publication of the work, generally the year of the scheduled product release

  See also "Copyright Date for a Revised Document" on page 187.
- The name of the owner of the copyright followed by "All rights reserved."

For example, any of these copyright notices are acceptable:

Copyright © 2009 PlirgSoft, Inc. All rights reserved.
Copyright 2009 PlirgSoft, Inc. All rights reserved.
© 2009 PlirgSoft, Inc. All rights reserved.

Your legal department might have a specifically approved copyright statement. The copyright portion of this statement should not be modified (except for the date) without consulting your legal department. The following example is a typical copyright statement.

Copyright © PlirgSoft, Inc., 1955 Nesfa Street, North Bay Village, Florida 33141-4332 USA. All rights reserved.

> This product and related documentation are protected by copyright and are distributed under licenses restricting their use, copying, distribution, and decompilation. No part of this product or related documentation may be reproduced in any form by any means without prior written authorization of PlirgSoft, Inc., and its licensors, if any.

If the work contains any third-party copyright notices, do *not* remove them. Only the lawful copyright owner may remove a copyright notice once it is affixed.

Copyright Date for a Revised Document

A previously copyrighted document that is recast, transformed, or adapted is considered a "derivative work" for purposes of copyright. For example, a document is a derivative work if you add a new section, chapter, or appendix or if you change the content.

If a document is a derivative work, change the date in the copyright notice to the current date. The current date is the year in which the derivative work is scheduled to be released. Your legal counsel might require the inclusion of the earlier dates in a copyright notice for a derivative work.

A document is *not* a derivative work if the material contains the following types of changes:

- Fixing incorrect spelling or style errors
- Changing the order of chapters
- Applying new templates

If a document is not a derivative work, use the original copyright date. Do not change the copyright date or add other dates.

Third-Party Copyrighted Information

If you are using third-party copyrighted material, follow these guidelines:

- Do not remove third-party copyright notices.

  Only the lawful copyright owner may remove a copyright notice once it is affixed.
- If you are paraphrasing information from other publications, use a regular citation, such as a footnote.

  You do not need to obtain permission from the information source.

  "Endnotes, Footnotes, and Bibliographies" on page 79 explains how to cite information from other sources.
- If you are reproducing exact tables or exact graphs or are using several paragraphs verbatim, you *must* obtain permission from the information source.

  Contact your legal counsel for details.

Protecting Trademarks

The scope and strength of a company's exclusive rights in its trademarks are weakened when the trademarks are not used properly. This precept applies even when the trademarks are registered. A number of well-known names, such as "escalator," "aspirin," and "cellophane," were once trademarks. Because these names have fallen into common usage and no longer identify a product from a single, specific company, the terms are now generic and can be used by anyone.

Trademark Terms

The following terms are associated with trademarks and are defined as follows:

- **Trademark legend.** Trademark attributions, without trademark symbols, for all trademarks that are marked in a particular document.
- **Trade name.** The name of a company, or its abbreviation, under which the company conducts business.

  Most companies do not place a trademark symbol after their trade names. Some trade names, though, are also used as trademarks. For example, Sun is the trade name of a company (Sun Microsystems, Inc.), but Sun™ is also the trademark for some of that company's products. Make sure that trade names that are also trademarks are used correctly. Trade names refer to companies, while trademarks distinguish the products and services that companies provide.
- **Service mark.** An identifier such as a word, phrase, logo, symbol, color, sound, or smell that a business uses to identify a service and distinguish it from similar services from its competitors. (For example, the famous MGM lion's roar is a service mark.)

Proper Use of Trademarks

To protect trademarks, you must have an understanding of the following topics:

- Trademark symbols
- Trademark symbol placement
- Trademark usage
- Trademarks and appropriate nouns

Trademark Symbols

The guidelines in this section pertain to print documentation. If you work on web pages, product documents on the web, or online help, see "Proper Use of Trademarks in Online Works" on page 192. In addition, determine which of the following guidelines are applicable to your work and use the guidelines accordingly.

Follow these guidelines when using trademark symbols:

- Use the appropriate symbol (™ or ®or ℠).

  If your authoring environment does not include one or more of these symbols, do the following:

  - If your authoring tool has superscript capabilities, type TM or SM or (R) and apply the superscript character format.

    Depending on your tool, you might also have to lower the point size. Try to match the point size to the size of any related symbols.

  - If your authoring tool does not have superscript capabilities, enclose the TM, SM, or R in parentheses, for example, “Plirg(TM) workstation.”

Trademark Symbol Placement

The guidelines in this section pertain to print documentation. If you work on web pages, product documents on the web, or online help, see “Proper Use of Trademarks in Online Works” on page 192. In addition, determine which of the following guidelines are applicable to your work and use the guidelines accordingly.

Follow these guidelines to ensure the correct placement of trademark symbols:

- Use the appropriate symbol to designate a trademark at the first occurrence and most prominent use of the trademark, which might be in the same place or in a different place:
  - On book spines and book covers
  - On title pages, in chapter titles and appendix titles, and in headings, depending on your organization's policy
  - In text

    The first use in text, which might also be the most prominent use, can be in the preface, a chapter, an appendix, or similar text element. If a document is lengthy, however, you can choose to repeat the symbol in any chapter or similar text element that contains many occurrences of the trademark.

  Note – Do not place a symbol after a trademark in captions, tables, figures, footnotes, or trademark legends. However, if a trademark appears only in a table, include the symbol in the table.

- If the same trademark appears in a document in two or more product names, use the trademark symbol only at the first occurrence of the trademark in text.

  For example, “The iPlirg™ 76 server, the iPlirg 769 server, and the iPlirg 5769 server were delivered to the customer today.” Even though the iPlirg trademark appears three times, use the trademark symbol only at the first occurrence of the trademark.

- If a trademark and service mark have the same name and both types of marks appear in the same document, one mark takes precedence.

Use the appropriate symbol only at the most prominent use and the first occurrence of the trademark *or* service mark in text.

For example, "Plirg offers the PlirgSoft℠ Manager Central service with the PlirgSoft operating environment. In this instance, "PlirgSoft" is used first as a service mark. Thus, you do not have to use the trademark symbol when PlirgSoft is used as a trademark in the same document. However, try to do the following:

- If the document primarily describes a product, write so that the ™ symbol is prominent and appears first.
- If the document primarily describes a service, write so that the ℠ symbol is prominent and appears first.

Trademark Usage

To ensure that trademarks are used correctly, follow these guidelines:

- Use trademarks as adjectives, not as nouns or verbs.

  Incorrect: UNIX is fun and easy to use.
  Correct: The UNIX® operating system is fun and easy to use.

  Incorrect: Plirgize the system.
  Correct: Enhance the system for the Plirg™ platform.

  For more information, see "Trademarks and Appropriate Nouns" on page 191.

- Do not use trademarks in the possessive or the plural, rather form the possessive or plural from the appropriate noun that the trademark describes.

  Incorrect: My dog ate the Macintosh's microphone.
  Correct: My dog ate the Macintosh™ system's microphone.

  Incorrect: Turn off your Scitex.
  Correct: Turn off your Scitex™ printer.

- Do not hyphenate trademarks.

  Incorrect: Use the PlirgPak-based application.
  Correct: Use the application based on the PlirgPak™ technology.

- Do not change the typeface of trademarks from the typeface of the surrounding text.
- Do not parenthetically define acronyms and abbreviations that are trademarks.
- Do not abbreviate trademarks.
- If a trademark is long or you work with space-constrained materials such as graphics or slides, you can define and use a short form that includes words to the right of the trademark symbol.

For example, you can write, "The Plirg Easy Write™ software (Easy Write) is ..." In subsequent text, you can use the short form if the reader can clearly ascertain that you are referring to that product. For example:

- Use the Easy Write software to ...
- The Easy Write product is ...
- Plirg Easy Write 2.0 is ...
- Easy Write is ...

If you mention two or more closely related products in the same document, you must clarify which product you are talking about at any given time.

Trademarks and Appropriate Nouns

Trademarks are proper adjectives. As such, they modify nouns. The term "common noun" (sometimes called "generic noun") refers to the noun that a trademark describes. Associate a trademark with an appropriate common noun. Do not capitalize the common noun.

Follow these guidelines when working with appropriate nouns:

- Use appropriate nouns with trademarks.

  The noun does not always have to follow the trademark. However, you must write clearly so readers know which trademark is used with the particular product, technology, service, or program described by the noun.

  For example, "Plirg recently enhanced these technologies: Easy Write,™ iPlirg,™ and PlirgPak™." The noun "technologies" is used with the Easy Write, iPlirg, and PlirgPak trademarks.

- Do not capitalize appropriate nouns unless the nouns are part of the name of the product or service.
- Do not use nouns such as "developer" or "vision" with trademarks or service marks because these terms are not products or services that your company provides.

  For example, you would not write "PlirgSoft developer," "PlirgSoft software developer," or "developer of PlirgSoft software." These phrases give the impression that a developer is the owner of PlirgSoft software. In this example, PlirgSoft software comes only from Plirg. PlirgSoft is a trademark of Plirg, Inc., not of a software developer.

  However, you *can* create a "defined term" for a trademark and then use the defined term throughout the document. For example, "Developers that write to the PlirgSoft platform ('PlirgSoft developers') use code examples as a main resource." "PlirgSoft developers" is now a defined term that you can use in the document.

- If you use an appropriate noun *in place of* a trademark, choose a noun that does not conflict with related terms in the document.

  Choose the appropriate noun, indicate the full name of the product or service associated with the noun (often, if helpful), and use the noun consistently.

  For example, sometimes the phrase "the server" is a useful replacement for "the iPlirg 5769 server." However, you cannot use just "the server" when comparing the iPlirg 5769 server with another server. Similarly, be careful when using words like "application" or "toolkit."

The following list provides some examples of frequently used common nouns in the computer industry:

application
architecture
client/server system
distributed computing solution
environment
equipment
features
files
graphical user interface (GUI)
hardware
interface
kernel
machine
operating environment
operating system
package
peripheral
platform
printer
program
screen
server
software
system
system software
technology
tool
unit
workstation

Proper Use of Trademarks in Online Works

This section provides trademark usage guidelines for the following types of online works:

- Web pages
- Product documents on the web
- Online help

This information supplements the basic trademark guidelines in "Proper Use of Trademarks" on page 188. If you have questions about trademark usage in online works, check with your legal counsel.

Trademark Usage on Web Pages

Consider establishing a central, publicly accessible site containing trademark information and a list of trademarks. You can then reference this site in the footer of each company web page. This method avoids the necessity for a trademark legend for every document. You also do not have to use symbols with trademarks or service marks in text.

Trademark Usage in Product Documents on the Web

Most product documents on the web can be downloaded, for example, in PDF. In this case, use the appropriate trademark symbol at the most prominent use of a trademark, which is usually in the title of the document. Also use the symbol at the first use of the trademark in text. If you want to use the symbol after other occurrences of the trademark in running text, you can.

Even if the product document on the web cannot be downloaded, all applicable trademarks and logos that appear in the document could be listed in the trademark legend, as is done with print documents, and placed in the legal notice for the product document.

Trademark Usage in Online Help

If users can download the online help, for example, in PDF, use the appropriate trademark symbol at the most prominent use of a trademark, which is usually in the title of the opening page of help. Also use the symbol at the first use of the trademark in text, as determined by the table of contents. If you want to use the symbol after other occurrences of the trademark in running text, you can.

If users cannot download the online help, you might need to use a symbol only after *registered trademarks* (not trademarks or service marks) at the first use in text, and in the title of an opening page of help, as applicable.

Even if the help cannot be downloaded, all applicable trademarks and logos that appear in the help could be listed in the trademark legend, as is done with print documents. This legend is could be placed in the legal notice for the online help or in the legal notice for the online product that the help supports.

Proper Use of Third-Party Trademarks

Treat "third-party" trademarks (trademarks from other companies) with the same respect as trademarks from your own company. Use them as proper adjectives with the correct , , or notices, and give them appropriate attribution as trademarks.

Follow these guidelines when using third-party trademarks:

- Mark third-party trademarks with the correct or notice on book covers and the first time they appear in text, including chapters, appendixes, and the preface. Most publications do not put trademark symbols in the table of contents, chapter or appendix titles, section heads, tables, or captions.
- Add the third-party attribution to the trademark legend of the legal notice.
- Make an effort to consult product or marketing groups, or the third parties themselves, to make sure that third-party trademarks are given appropriate attributions. Corporate web sites often provide such information. If you cannot obtain the information, include a general attribution similar to the following example in the trademark legend after the specific attributions:

  All other product names mentioned herein are the trademarks of their respective owners.
- Do not hyphenate third-party trademarks.
- Only put trademarks of different companies next to each other if a license or agreement exists between the two companies.

  For example, you would not write "Plirg Ada" unless a licensing agreement existed between Plirg, Inc., and the U.S. Department of Defense.
- Do not parenthetically define existing acronyms or abbreviations that are trademarked terms.

Referencing External Web Sites

This section provides guidelines for referring to third-party web sites.

Pointing to Third-Party Web Sites

A third-party web site might contain information that you want to reference in a document. Choosing and referencing an appropriate third-party site might involve the following:

- Determining which third-party site to reference
- Determining which third-party URL to use
- Adding a disclaimer and any required third-party wording
- Preventing unapproved references to third-party sites

Note – The guidelines in this section pertain to pointing to third-party sites in printed documentation. If you work on web pages, product documents on the web, or online help, determine which of the following guidelines are applicable to your work and use the guidelines accordingly. Considerations can include your staffing resources and whether to point to nice-to-know information. Also, consider the length of time the material might be available online and whether technical input is required.

Determining Which Third-Party Site to Reference

Sometimes, a third-party web site might be the best source of information for a topic. If you want to point to a third-party site, first find out whether the site prohibits such use.

Often, a third-party site has a Terms of Use page that is available from its home page or legal notice page. The Terms of Use page likely states the third-party's policy for pointing to the site's contents.

- If the site clearly states that you cannot point to its contents, do not point to that site.
- If you are not sure whether you can point to the site's contents, contact your legal counsel.

When you know that you can legally point to a particular third-party site, ask engineering staff or subject matter experts to review the material. The reviewers can determine the value and accuracy of the material and might even be able to provide a better resource. Allow extra time in your documentation schedule to gather this legal information and technical data.

Determining Which Third-Party URL to Use

After determining which third-party site to use as a source of information, point to a "safe," pertinent, and easily accessible URL.

- Check the third-party Terms of Use page for any restrictions on pointing to a particular URL.
  - If the third-party terms of use only allow pointing to the home page, point to the home page.
  - If the third-party terms of use do not prohibit pointing to a page other than the home page, you can point to another page.

  If many links are needed to reach the required information, consider using another source.
- Point to need-to-know information, not to information that is nice to know.

  For example, you might point developers to a standards site, such as `http://www.ietf.org/rfc`. Do not direct users to a general, nice-to-know site on computer literacy, for example, or to information that essentially duplicates other content.
- Decide whether users might have difficulty accessing the information or might lose patience waiting for information to be displayed.

  For example, does the site take a long time to load due to large graphics files? Does the site require plug-ins or other software that users might have to download? Has the site crashed on any occasions when you have tried to access it?
- Use text to point to a third-party site.

  For guidelines on referencing a URL, see "Referencing URLs" on page 273.

  Your authoring tool might enable you to use "link text," which is text that you want to appear as the cross-reference. Clicking the link text takes users to the URL that you also specify when creating the link text.
- Link to a third-party site by clicking on a logo or other image *only* if the site's terms of use expressly permit the use of the logo or image.

Adding a Disclaimer and Any Required Third-Party Wording

Include the following elements in any document that points to third-party sites:

- A disclaimer similar to the following example, which appears as a Note:

  Note – Plirg is not responsible for the availability of third-party web sites mentioned in this document. Plirg does not endorse and is not responsible or liable for any content, advertising, products, or other materials that are available on or through such sites or resources. Plirg will not be responsible or liable for any actual or alleged damage or loss caused or alleged to be caused by or in connection with use of or reliance on any such content, goods, or services that are available on or through such sites or resources.

- Any special wording if, and as, mandated by the third-party site's terms of use

Preventing Unapproved References to Third-Party Sites

When referring to third-party web sites, use caution:

- Do not endorse or criticize a book or other content that you reference.

  Be straightforward and neutral. For example, "For information about UNIX commands, see `http://www.unix4fun.com`."
- Do not use any type of framing or inline linking.

  The page that you link to cannot appear in the browser simultaneously with the page from which you are linking.
- You might not want to point to a competitor's site or documentation.

Protecting Confidential Information

Confidential information is defined as any information that is not commonly known and that when mismanaged, compromised, and ultimately disclosed to unauthorized parties could be detrimental to your company's competitive advantage or could adversely affect your company's operation of business. This information might be technical, financial, business strategic, or operational.

Because this information is not commonly known to others, the information provides your company with a financial advantage in the marketplace. This information has a property value to your company.

Generally, these forms of confidential information are safeguarded as trade secrets. A *trade secret* can be any information, technical or nontechnical, that is considered confidential. Do not release trade secret information to competitors or the public domain.

This section discusses the types of information that are confidential and provides guidelines for correctly identifying and labeling confidential information.

- Identifying confidential information
- Classifying and labeling confidential documents
- Classifying and labeling confidential electronic communication
- Protecting confidential information in examples

Identifying Confidential Information

Consider information to be confidential if any of the following statements are true:

- Competitors should not have unrestricted or unauthorized access to the information.
- The media should not have unrestricted or unauthorized access to the information.
- Your company could lose potential copyright protection if protected material, such as software or technical manuals, is released without proper copyright notification.
- Your company could lose the ability to protect the information by patent or trade secret status if the information becomes part of the public domain.

The following types of information are confidential, whether written or in electronic or digital form:

- Drafts of manuals, white papers, and product notes
- Technical data
    - Object code
    - Source code and source documents
    - Flow diagrams
    - Schematics
    - Host names of computers at your company
    - Public IP addresses
    - Domain names
- Business data and operations
    - Financial results (except publicly published results)
    - Merger and acquisition activity
    - Alliance negotiations
    - Workforce information
    - Purchasing and bid data
    - Marketing information and strategies
    - Sales forecasts
    - Customer lists and profiles
    - Nonpublic strategies and ideas

- Detailed information about new products before public announcement
  - Development information such as project code names and descriptions
  - Product features (processing speed, graphics capability)
  - Project timelines and target dates (beta, final release)
  - Market placement and strategies, and customer information
  - Costs, budget details, pricing, and other financial information
  - Bugs and restrictions
  - Design, diagnostic, and reliability data
- Email messages, electronic files and media, and web pages of a confidential nature
- Computer printouts, spreadsheets, and status reports of a confidential nature
- Presentation, marketing, and educational materials, including handouts, transparencies, slides, video tapes, and other media of a confidential nature
- Employee personnel information, such as performance reviews and salary information, as well as employee names and telephone numbers

Classifying and Labeling Confidential Documents

Confidential documents are not limited to only processor-generated text or spreadsheet documents. Confidential documents can also include plot drawings, schematics, handwritten notes, brochures, overhead transparencies, and microfiche, that is, virtually any information in any tangible form. Confidential documents must be identified as such through appropriate classification and corresponding labeling on each printed page. This labeling must be present from the time the documents are created until the documents are released to the public or are securely destroyed.

For confidential electronic communication, such as email, web sites, and other media, read this section for descriptions of the most commonly used confidential classifications and labels. For specific guidelines, see "Classifying and Labeling Confidential Electronic Communication" on page 200.

Note – *Incorrect* classification of information can also jeopardize information protection. If you cannot determine whether information needs to be labeled as being confidential or if you are not sure which label to use, contact your legal counsel. Also, your company might use different classifications than those referenced in this chapter.

This section explains the three most commonly used confidential classifications. These classifications are often preceded by the name of the company (for example, "Plirg Confidential: Internal Only"). If your authoring tool can produce bold type, use bold with these classifications.

- **Confidential**

  This classification is for information that is confidential to your company. Many companies restrict access to information marked Confidential to regular employees, contractors, and non-company third parties who have signed a confidential nondisclosure agreement. When determining whether information should be classified "Confidential," consider the following:

  - Value and sensitivity of the information
  - Potential consequences to your company if the information is inappropriately disclosed
  - Number and type of people who need access to the information
  - Nonpublic nature of the information
  - Timeliness of the information

  Use this label for general "need to know" information for the purpose of meeting the business needs of a defined and limited group. This disclosure group may be as large as all employees, for example, notifying employees about job listings throughout the company, or narrowly defined, for example, "Confidential: Marketing Department Only." A few examples of confidential information that might warrant specific limited disclosure within your company include product design information, project timelines, budget details, and status reports of a confidential nature.

  Note that some confidential information may need to be unclassified at some point. If such information is released to the public, remove any confidential markings, and make sure that any copyright date reflects the year of disclosure to the public.

- **Confidential: Internal Only**

  Use this label for all prerelease product documentation and information, such as manuals, release notes, white papers, and specifications, which are distributed to product teams. *Remove* the label before producing the final version of the document *unless* the document remains confidential. For example, do not remove the confidential label from certain source code documents. If a source code document is released to the public, remove any confidential markings, and make sure that the copyright date reflects the year of disclosure to the public.

- **Confidential: Registered**

  Use this label for highly sensitive information that uses registration controls such as document numbering, document assignment logging and tracking, and copy controls.

Classifying and Labeling Confidential Electronic Communication

Information that is shared in an electronic format can seem less tangible than a piece of paper. If the information is considered confidential, the sensitivity of the information remains the same and must be treated the same as a printed document.

To protect electronic communication, follow these guidelines:

- **Email, files, and directories.** Identify and label "Confidential" or "Confidential: Internal Only" (*not* "Confidential: Registered") email, files, and directories.

  Put the label in the header of the email or centered at the start of the message or text.

  Confirm the names on an alias before sending information to a large audience. Create smaller aliases for particularly sensitive topics. Do not distribute messages beyond the alias.

- **External web sites.** Do not post confidential information to external web sites.

  Information classified as confidential should not be placed on or transmitted over the Internet or other public network.

- **Other media.** Appropriately classify and label audiovisual forms of information that contain confidential material. Label the electronic format, containers, and packaging. Audio and video tapes and CD and DVD discs should contain the classification label on the leader or at the beginning of the recording.

Protecting Confidential Information in Examples

If you create sample text, files, screen captures, illustrations, or any other types of examples in documentation, you must protect certain types of information. This information can include names of computers, or IP addresses, or possibly entire network domains. Code names, URLs, and public names, addresses, and telephone numbers must also be protected. Follow these guidelines:

- Do not use real host names of computers at your company in documentation.

  Suppose you create files with real computer names to test the software being documented. If you want to include these files in a document, you must change the real host names to fictitious names. Try to create culturally neutral, easily translated host names, such as `host1`, `newhost`, or `myhost`.

- Do not use public IP addresses in documentation. Different numbers should be used for the most common types of IP addresses: IPv4 and IPv6.

  - **IPv4 addresses.** The Internet Assigned Numbers Authority (IANA) has set aside three blocks of the IPv4 address space for use in corporate intranets.

    The IANA's numbering scheme enables companies to use private, "internal" IP addresses inside a firewall. Because these private IP addresses cannot be used by the public, their use in documentation does not conflict with IP addresses used on the Internet.

The following table shows the *range* of numbers that the IANA has set aside for internal IPv4 addresses. The numeral 255 is the highest number that you can use in respective portions of the IP address. When you use IP addresses as examples in documentation, you *must* stay within this numbering scheme.

Network Class	Starting Number	Ending Number
A	`10.0.0.0`	`10.255.255.255`
B	`172.16.0.0`	`172.31.255.255`
C	`192.168.0.0`	`192.168.255.255`

For more information about these standards, specify the RFC number, RFC 1918, at the RFC repository for the Internet Engineering Task Force (IETF) web site at `http://www.ietf.org/rfc`.

- **IPv6 addresses.** The IETF has reserved the `2001:0db8::/32` prefix for use in documentation. This prefix represents the leftmost 32 bits of the address. Use the digits `2001:0db8` as the first two segments of a unicast IPv6 address, for example, `2001:0db8:3a4c:15::83:2a:33/64`.

  Unicast IPv6 addresses are based on IPv6 prefixes that are created by people. Link-local IPv6 addresses are created by the computer, based on its machine address. Link-local addresses start with the segment `fe80` and do not have to be protected legally in examples.

- Do not use real domain names in documentation.

  Instead, use these domain names in examples:

  - `example.com`
  - `example.net`
  - `example.org`

  Any subdomain can be combined with these domain names as appropriate, for example, `janepc.example.com`.

  For details about standard domain names, retrieve RFC 2606 at `http://www.ietf.org/rfc`.

- Do not use code names in examples.
- If you need to use a URL as an example in text, use your own company's external site.

  If a screen capture contains a URL that is not to be made public, change the URL and then take the screen capture.

- If you need to use a name, address, or telephone number of a person or company in an example, use a general, not readily identifiable name, address, or number.

  For example, instead of using your name, a friend's name, or another employee's name, use a generic name, such as John Smith or Susie Jones. A company name could be ABC Corporation. For an address, you could use 1234 Five Street, Anywhere, USA 12345. For a telephone number, use +1-*xxx*-555-*xxxx*, where *xxx* is any number and *xxxx* is a number other than 1212.

◆ ◆ ◆ CHAPTER 10

Types of Technical Documents

Many different types of documents exist. The documents range from a single-chapter manual, such as a white paper or simple installation guide, to a highly technical user's guide or training manual. Technical documents are often part of a documentation set that contains several documents. Many companies produce documents do not follow a book model.

This chapter discusses the following topics:

What Is a Documentation Set?

A *documentation set* includes one or more types of documents that a customer receives with a product. A typical documentation set might include the following:

- Manuals
- Textual pieces of a CD or DVD package
- Online help
- Release notes
- Other related documents, such as white papers

If you work at a large company, you might create or update a specific document, such as a user's guide or reference manual. In smaller documentation groups, a writer might work on all parts of the documentation set, including planning documents.

Documentation Plans

Publications groups often use planning documents to describe the contents and packaging of documentation sets and the individual documents within these sets. You can also use planning documents to describe individual documents. Your publications group probably uses standard planning documents or templates to describe new and revised documents and documentation sets.

Involvement in the creation of planning documents depends on the scope of the project and your position within the writing team.

Documentation Set Plan

The *documentation set plan* describes the overall characteristics of a proposed documentation set. You might think of a documentation set plan as a publications architecture document, similar to an engineering specification.

For large documentation sets, the publications manager or project lead often designs the documentation set plan with input from individual contributors. For smaller sets, individual contributors might provide documentation set plans.

Documentation Set Plan Template

Use a documentation set plan template when you create a new set of manuals. The plan template might include the following:

- Structure of the set
- Audience description
- Document content plan, for each document in the set
- Documentation schedule
- Reviewer list
- Roles and responsibilities of publications members and product team members
- Issues, including decisions that have not been made or known problems
- Dependencies that must be met for the project to be successful
- Localization plans
- Related documentation efforts
- Competitive analysis
- Publications quality assurance plan
- Media formats, such as print and online
- Format of the set, including bindings, cover type, and other items
- Packaging plans

Documentation Set Revision Plan Template

Use a documentation set revision plan template when you update one or more manuals that are part of a set or when you add or delete one or more manuals from a set. The plan template might include the following:

- Structure and delivery, including differences from the previous set
- Documentation schedule
- Reviewer list

Document Plan

A *document plan* describes the characteristics of a *specific manual*. The plan might include a content outline, production considerations, and an explanation about how an individual document will be implemented. Typically, the writer or project lead creates the document plan.

Document Plan Template

Use a document plan template when you create a new, stand-alone manual that will *not* be part of a documentation set. The document plan template might include the following:

- Content outline
- Audience description
- Manual schedule
- Reviewer list
- Roles and responsibilities of publications members and product team members
- Issues, including decisions that have not been made or known problems
- Dependencies that must be met for the manual to be effective
- Manuals that will be referenced from this document
- Localization plans
- Publications quality assurance plan
- Media formats, such as print and online
- Format of the manual, including size, cover type, and similar details
- Packaging plans
- Related documents
- Competitive analysis

Document Revision Plan Template

Use a document revision plan template when you revise an existing stand-alone manual that is not part of a documentation set. The template might include the following:

- Content outline
- Manual schedule
- Reviewer list
- Related documentation
- Structure and delivery of the revised manual

Abstracts

An *abstract* conveys to potential readers what is in a book so that they can make an informed decision about whether to read the book. When writing an abstract, reduce text to a few sentences, and to no more than one or two short paragraphs. Use the appropriate trademark symbols where necessary.

Structure of Manuals

Manuals might consist of either a single chapter or multiple chapters.

Manuals With a Single Chapter

Manuals with a single chapter are usually small and narrowly focused on a single subject. Examples of manuals with a single chapter include the following:

- Simple installation manuals
- Release notes
- Product notes
- White papers
- Documentation roadmap

Note – Release notes and product notes require special consideration. See "Release Notes and Product Notes" on page 213 for more information.

The following table lists the components of a typical single-chapter manual. The components are listed in the order in which they usually appear within the manual. The *optional* components can be provided at your discretion. For example, you probably do not need a table of contents if the document has only one or two pages. Likewise, an index is unnecessary for a document of fewer than 20 pages. A manuals with a single chapter usually requires a title page and a legal notice page.

TABLE 10–1 Possible Components of Manuals With a Single Chapter

Part of the Manual	Requirement
Title page	Usually required.
Legal notice	Usually required.
Table of contents	Optional.
List of figures	Optional. Helpful if many numbered figures are used.
List of tables	Optional. Helpful if many numbered tables are used.

TABLE 10–1 Possible Components of Manuals With a Single Chapter *(Continued)*

Part of the Manual	Requirement
List of examples	Optional. Helpful if many numbered examples are used.
Chapter	Required.
Appendix	Optional.
Index	Recommended if the chapter has more than 20 pages.

Manuals With Multiple Chapters

Manuals with multiple chapters are the most common types of technical manuals. These manuals require a title page and front matter.

In addition to being divided into chapters, some manuals are further divided into parts. A part contains one or more chapters. Manuals can be divided into parts for various reasons. One example is a single manual that is both a user's guide and a reference manual.

If a manual is divided into parts, the manual must have at least two parts. The parts should also be identified by a part divider page. See "Part Dividers" on page 210.

The following table lists the components of manuals with multiple chapters. The components are listed in the order in which they appear within a manual. *Optional* means that the component of the manual might be required in certain writing groups. When in doubt, check with your writing manager or project lead.

TABLE 10–2 Possible Components of Manuals With Multiple Chapters

Part of the Manual	Requirement
Title page	Required.
Legal notice	Required.
Table of contents	Required.
List of figures	Optional. Helpful if many numbered figures are used.
List of tables	Optional. Helpful if many numbered tables are used.
List of examples	Optional. Helpful if many numbered examples are used.
Preface	Recommended.
Part I	Optional. Use a Roman numeral.
Chapter table of contents	Optional.
Chapters	Required.
Part II	Required if a Part I is used.
Parts III and higher	Optional.
Appendixes	Optional.
Glossary	Optional.

TABLE 10–2 Possible Components of Manuals With Multiple Chapters (*Continued*)

Part of the Manual	Requirement
Bibliography	Optional.
Index	Recommended for manuals that are longer than 20 pages.
Revision history	Optional.

Descriptions of the Manual Parts

This section describes the general editorial formats of the parts of a manual. The parts are listed in the order in which they appear within a manual.

Title Page

Most manuals have a title page, which usually contains the manual title, current address block for your company, and release date. Your company's specific information might also include elements such as the corporate logo, corporate telephone number, and document revision information.

Legal Notice

The legal notice contains the various copyright and trademark statements.

Table of Contents

The table of contents can list the first-level headings, second-level headings, and third-level headings in a manual. Headings can be numbered or unnumbered, depending on the style determined for your document.

A manual should have a table of contents if the manual has more than one chapter. The table of contents usually begins on a right page (an odd-numbered page) in a printed manual.

List of Figures

The list of figures lists all numbered figures in the manual. Consider including the list of figures if the manual has more than one numbered figure and more than one chapter. The list of figures usually begins on a right page in a printed manual.

List of Tables

The list of tables lists all numbered tables in the manual. Consider including the list of tables if the manual has more than one numbered table and more than one chapter. The list of tables usually begins on a right page in a printed manual.

List of Examples

The list of examples lists all numbered examples in the manual. Consider including the list of examples if the manual has more than one numbered example and more than one chapter. The list of examples usually begins on a right page in a printed manual.

Preface

The preface describes the purpose and scope of the manual and includes an overview of the content of the manual. The preface usually begins on a right page in a printed manual.

Depending on the purpose and complexity of the manual, the preface might contain all or some of the sections listed here. The preface might also contain sections specific to a documentation set.

First, explain the purpose of the manual in one or two sentences. Next, include sections similar to the following if the additional information is relevant to the book:

- **Who Should Use This Book.** Describe the audience or class of reader for whom the manual is intended. The audience description might include the following information:
  - Required knowledge, such as a specific programming language
  - Required experience or familiarity with specific software or a hardware platform
  - Definition of the type of user or functional responsibility, such as applications programmer, system administrator, or field engineer
  - Tasks the user might perform using this product
- **Before You Read This Book.** If the user is required to read other documents before effectively using this manual, list those other documents.
- **How This Book Is Organized.** Briefly describe the contents of the manual. Consider using a active cross-references to list each chapter and appendix.
- **Related Books.** List titles of internal documents that are related to the manual. Also list third-party books, and their authors and publishers, that are mentioned in the text or that readers might find useful.
- **Accessing Documentation Online.** Inform readers how to view documents online if your company provides this service.
- **Typographic Conventions.** List and explain special symbols, characters, or typography that are used in the manual.
- *CompanyName* **Welcomes Your Comments.** Provide contact information for users to give feedback about the manual if you have a feedback mechanism in place.

Chapters

Chapters comprise the main body of a manual. A manual must contain at least one chapter.

The three general types of chapter page numbering and heading styles are:

- **Single.** Single chapters are not numbered. There is no chapter number, first-level headings are not numbered, and the page numbers run sequentially starting from 1. This format is used only in manuals with one chapter. See "Manuals With a Single Chapter" on page 206.
- **Multiple, unnumbered.** The chapters are numbered but the headings are not. Pages are numbered sequentially, starting from 1, or pages can be numbered sequentially within each chapter, starting from 1–1.
- **Multiple, numbered.** The chapters are numbered, the first-level headings through the third-level headings are numbered, starting from 1.1. Pages are numbered sequentially within each chapter, starting from 1–1.

The type of page numbering style that you use is optional and depends on various factors such as personal preference, history of the manual, or the manual's subject matter. All chapters begin on a right page in a printed manual.

Part Dividers

A lengthy book might be divided into parts that require *part dividers* to group similar chapters in the document. Part dividers are numbered with Roman numerals, for example, Part I, Part II, Part III, and so on, and include a title. The part dividers can provide a cross-reference listing of the chapters that are included in the part. You might also include text and illustrations, either on the front or on both the front and back of the part divider.

Never include any information on the divider that is essential to the topics that are discussed in the chapters. For example, instructions to turn off power before proceeding with tasks are not appropriate on the divider.

A part divider appears on a right page in a printed manual and does not include a page number.

Appendixes

Appendixes provide supplementary information to the main body of the manual. Appendixes appear at the end of the manual. Typical material for appendixes includes the following:

- Long programming code examples
- Long lists, charts, and tables
- Summary information, such as a list of a program's function keys
- Technical specifications

Do not use an appendix as a repository for miscellaneous information that you are unable to integrate in the body of the text.

Appendixes can use numbered or unnumbered section headings, depending on the chapter style of the manual. For example, you would use numbered section headings in appendixes for manuals that use numbered section headings in chapters.

Each appendix is designated by a letter, for example, Appendix A, Appendix B, and so on. If the book contains a single appendix and your authoring tool enables you to change the setting, the name of the single appendix should be Appendix rather than Appendix A. Each appendix begins on a right page in a printed manual.

Glossary

A glossary is an alphabetical list of defined terms, phrases, abbreviations, and acronyms. The glossary defines terms that might not be clear to the reader.

The glossary is included within a manual in the following location:

- At the end of the manual, following the last appendix
- At the end of the manual, following the last chapter if the manual contains no appendix
- Before a bibliography or index

The glossary begins on a right page in a printed manual.

Your publications department might also have developed a glossary with more specific definitions for terms that are specific to your product area. For more information, see Chapter 17, "Glossary Guidelines."

Bibliography

A bibliography is a list of the sources to which you refer in your manual. Bibliographies follow a specific format for each kind of resource that is cited. A bibliography begins on a right page in a printed manual. For more information, see "Writing Bibliographies" on page 80.

Index

An index is usually recommended for a manual of more than 20 pages. Indexes begin on a right page in a printed manual. For more information, see Chapter 18, "Indexing."

Types of Hardware Manuals

This section briefly describes some of the types of manuals that are typically written for a hardware product. Hardware documentation can include the following types of manuals:

- **Installation guides.** Include system installation guides for workstations, servers, and external expansion systems, as well as internal installation guides for field-replaceable units (FRUs) inside systems. The FRUs could include subsystems (mass storage systems and switches), individual devices (drives), components (CPU modules and memory modules), and chassis units (fan trays).
- **System overview guides.** Cover technical features and functions of the system for users who set up and administer large server systems.

- **User's guides.** Usually contain installation, configuration, and setup instructions, as well as information about administration and troubleshooting for users who set up and administer systems.
- **Service manuals.** Are typically prepared for platforms and storage systemsrather than for a single board or a single device. Many large companies have internal service organizations that use the service manuals to maintain their equipment. Therefore, the readers of service manuals might be experienced service technicians who work on a given company's equipment only occasionally. When writing a service manual, you cannot make any assumptions about the reader's knowledge of your company's equipment.
- **Configuration guides.** Are typically written for families of products. These manuals typically tell the user how products fit together and how to add certain devices to the system configuration. Your Product Marketing staff might use this type of manual to assist customers in meeting their installation requirements.

Other hardware manuals can include these types of guides: specification, site planning, diagnostics, getting started, and safety and compliance.

Types of Software Manuals

This section briefly describes some of the types of manuals that you might write for a software product. Software documentation can include the following types of manuals:

- **Installation guides.** Explain how to install and configure the software on the user's system.
- **Programmer's guides.** Vary in scope, depending on whether you are discussing the programming features of the operating environment, a language compiler application, or an application programming interface (API). Typical operating system programmer's guides might range from an application packaging developer's guide, which is written for a general audience of application developers, to a book about writing device drivers, which is directed to highly specialized system programmers.

  Programmer's guides explain how to write programs that take advantage of the features of a particular product. Compilers and language-development tools have their own documentation sets that are specifically geared toward programmers who use the tools.
- **System administration guides.** Explain how to set up, maintain, and troubleshoot software on computers and networks of computers.
- **User's guides.** Explain how to use the features of a software product. The guide's complexity depends on how sophisticated readers must be to use the product.
- **Reference guides.** Provide very specific details about a software product, such as syntax statements for programming languages or explanations of each feature of a graphics software package. The subjects are usually arranged in alphabetical order.

Release Notes and Product Notes

Release notes for software products and product notes for hardware products are the final documents that are written before a product ships. The purpose of these notes is to inform the customer about any major problems with hardware, software, or documentation that were discovered after the manuals were printed. Release notes also contain other late-breaking information, and additional information needed at installation.

Release notes might be printed material in product boxes or in ASCII or HTML on the product CD or DVD. Release notes might also become part of a product's online documentation set if one exists.

Typically, release notes go to production for a "quick print" one to two weeks before the product is shipped. Release notes and product notes should be as short and succinct as possible.

Release notes might include the following sections:

- Title page
- Legal notice
- Table of Contents
- Preface
- Installation Bugs
- Runtime Bugs
- New Features
- End of Support Statements
- Driver Updates
- Patches and CERT Advisories
- Documentation Issues

Product notes might include the following sections:

- Title page
- Legal notice
- Table of Contents
- Preface
- Getting Help
- Compatibility
- Known Problems With the Hardware
- Known Problems With the Software
- Documentation Issues
- Bugs Fixed Since the Last Release

Other Product Documents

Though manuals form the bulk of typical technical documentation, you might also find yourself writing other types of documents. Some of these documents could be considered manuals, but their uniqueness demands special consideration.

- **White papers.** Optional documents that are usually used to publicize a new product or a new feature before official documentation is available. The involvement of a project's technical writer ensures that all product information is consistent with the product documentation, uses the standard documentation format, and meets editorial guidelines.

  When assigned to edit or write a white paper, you might interact with Product Marketing, Engineering, or other groups regarding schedules, customer requirements, and other issues.
- **Online help.** Information that an application displays to assist users in performing tasks. Online help can appear as phrases or sentences with the user interface, or as more comprehensive information in a separate help window.
- **CD or DVD text.** Many software products are supplied on CD-ROM or DVD media. The textual pieces of the media package are often considered part of the documentation set, and typically consist of CD or DVD faceplate text and a text insert.
  - **CD or DVD faceplate text.** Appears on the disc faceplate and typically includes the following information:
    - Product title
    - Hardware platform supported
    - Operating system and other software compatibility
    - Applicable legal notice and third-party trademarks
    - File system format, such as Macintosh OS X
  - **Text insert.** A glossy brochure that forms the cover text for the CD or DVD package. The CD or DVD text insert typically consists of a title page, legal notice, and textual information. The text insert usually includes a brief description of the product and might include instructions for mounting the CD or DVD.
- **Demos.** Compiled, working examples that customers can try. The purpose of some demos is to provide customers with an opportunity to test that the software has been installed properly. Other demos enable customers to see and try new features.

  Demos are usually produced by engineers who write the code for the feature being demonstrated. You might be asked to review the demo, or to write one or more of the following documents:
  - README file
  - Instructions for loading and running the demo
  - An explanation of what the demo is intended to illustrate
- **Road maps.** Can provide the big picture perspective of a product and its components, or it can explain where to find the documentation and the possible order in which to access the documentation.

Training Documentation

Training documents are yet another type of technical documentation. Training documents vary widely in format, style, and delivery method, depending on how they are used. For example, a training document that an instructor uses in front of a classroom is different from a training document that a student accesses over the web for self-paced learning.

Typical training documents include the following:

- Training materials for use in instructor-led courses delivered in classrooms

  These materials might include a student guide, an instructor guide (consisting of the student guide plus additional notes), overheads, workbooks, and study guides.
- Training materials delivered on a CD or DVD that use proprietary software
- Training delivered over the web and that might use multimedia capabilities

Student Guides and Instructor Guides

Student guides and instructor guides are used in instructor-led training. This type of training typically takes place in a classroom with a group of students. The general editorial format of a student guide resembles the format of a hardware or software manual, except that the information is organized according to modules and learning objectives rather than by chapters.

Student guides might include the following components:

- Title page
- Legal notice
- Table of Contents
- Preface
- Modules
- Labs
- Exercises
- Appendixes
- Index

Instructor guides closely resemble student guides, except that the instructor guides contain additional directions and notes for the classroom instructor.

Other Training Documents

Other documents that are used for training purposes might include the following components:

- Study guides
- Student workbooks
- Lab manuals
- Certification exams
- Video or audio presentations
- Technology-based training (TBT) delivered on CD or DVD
- Classes accessed on the web
- Consulting materials, such as skills assessments or white papers

◆ ◆ ◆ CHAPTER 11

Working With an Editor

Writing computer documentation involves converting raw information from engineers and marketing professionals into a useful, well-written document. The final document often is a result of efforts from the entire publications team, including writer, editor, designer, illustrator, and production coordinator. However, the content of the document is most closely developed through the work of writer and editor.

This chapter discusses the following topics:

Technical Editor's Role

An editor helps a writer focus on content and effective presentation and provides another set of eyes to check all the details. The partnership of writer and editor produces easy-to-use, high-quality, effective documents.

Any editor is concerned with use of language, flow, tone, grammar, punctuation, capitalization, spelling, sentence structure, consistency, and so forth. However, a *technical editor* is also concerned with technical content, compatibility of the technical depth with the reader's background, and effective communication of technical information. Other areas of concern include consistent use of technical terms and symbols, and careful coordination of text and artwork. By marking text and suggesting an alternative, an editor indicates to a writer that the original might be, for example, misleading, awkward, imprecise, confusing, or incomplete.

Editor's Role in Producing Online Documents

Editors can help the writers of online documents with the following tasks:

- Identifying the document's readers and purpose.

  An editor can help a writer use this information to decide which documents to optimize for online, which documents to optimize for print, and how to establish priorities for conversion projects.

- Defining online document structure and the links that are under the writer's control.

  Writers have control over links that are embedded in the text and links that are in jump lists. However, writers typically do not have control over standard navigational aids such as Back, Forward, and Home, which are predefined in design templates.

- Ensuring that the document accommodates scanning and nonsequential access by readers.

  An editor can help ensure that text complies with online writing style guidelines. See Chapter 5, "Online Writing Style."

- Assisting with a visual inspection of the online document.

  The visual inspection checks for inconsistencies and formatting problems that result from document conversion.

- Verifying that all links are contextually appropriate.

  An editor can review the link wording and surrounding text to ensure that sufficient context minimizes reader disorientation.

- Verifying that links are appropriately placed within the document to avoid overlinking or underlinking.

- Assisting with usability testing.

  Usability testing can determine whether document navigation follows pathways that readers are likely to follow.

Types of Editing

A document could undergo more than one editorial review, each for a different purpose. The type of edit that a document receives usually depends on where the document is in the product cycle. For example, a *developmental edit* of a document occurs early in the cycle, around the pre-alpha test or alpha period when there might be more time to address issues such as organization and structure. A *copy edit* is best during the beta review, when the manual is more complete and stable. And finally, *proofreading* is the last review a document receives.

Developmental Editing

Developmental editing is hard to define because its functions depend on the documentation set or book under consideration. Think of developmental editing as a document production phase that assesses the document's overall focus and direction. This edit is the phase when a documentation set or a book is restructured, chapters or sections are reorganized, and major rewriting is done. The issues that the editor raises during a developmental edit can affect the character of subsequent sections or chapters of a document. This effectiveness is increased if the edit is done on a sample chapter or an early draft of a manual. Some global copy editing issues can be raised at this time as well, especially when these issues provide the writer with examples of style or word usage.

Developmental Editing Guidelines

Structure and Organization

- Audience definition, purpose of document, and how to use the book are clear.
- Information is appropriately presented for the audience.
- Document, and each chapter, accomplishes its stated purpose.
- Concepts flow logically.
- Superfluous or redundant material is eliminated.
- Headings are useful, descriptive, and specific.
- Information is easy to find.
- Information is task oriented where appropriate.
- Reference and conceptual information are eliminated from task descriptions.
- Distinctions between parts and chapters are clear.
- Page numbering scheme is appropriate for the type of book.

Writing

- Reader context is established and reinforced.
- Ideas are given appropriate weight in text.
- Tone is appropriate for the reader and to the focus of the book.
- Critical information is covered clearly.
- Task-oriented writing is clear. User actions and system actions are distinct.
- Assumptions are clearly supported.
- Writing and layout are optimized for online presentation.

Style

- Terms are used consistently and appropriately.
- Terms are defined and used in context correctly.
- Terms and abbreviations avoid jargon and follow guidelines for localization.
- Documentation set conventions are established and followed.

Formatting and Layout

- Document conforms to house publications standards.
- Standard templates and formats are used.

Illustrations

- Illustrations appear where needed.
- Artwork is integrated within the text.
- Tables, figures, and illustrations are used effectively and appropriately.
- Illustrations follow artwork and localization guidelines.

New Elements

- New graphics or presentation techniques are identified and used effectively.
- Innovations meet your company's design and usability standards.

Copy Editing

The editor does minimal rewriting, if any, during a copy edit. Issues regarding structure and organization are addressed throughout the developmental edit. At the copy editing stage, the editor does two kinds of review: *mechanical editing* and *editing for house style*. Mechanical editing addresses issues such as punctuation, capitalization, subject-verb agreement, and so forth. Editing for house style involves interpreting and applying your company's style guidelines. The editor also reads for correct usage of fonts, tags, or other markup, and for structural elements such as tables, illustrations, lists, and procedures. The best time for a copy edit, also called a *line edit*, is before or during the beta review.

Copy Editing Guidelines

Readability

- Sentences are clear, direct, and concise.
- Repetition is used effectively.
- Parallel structure is used effectively.

Style

- Headings, lists, and sentences have parallel construction.
- Headings are written as short, precise abstracts of their associated content, and they follow hierarchy guidelines.
- Voice and tone are consistent.

Transitions

- Text is easy to follow.
- Information is complete and appropriately placed.
- Transitions between parts, chapters, and sections are clear.
- Transitions are effective online and in hard copy.
- Cross-references are correct, worthwhile, and sufficient.

Grammar

- Sentences are complete.
- Subjects and verbs agree, and pronouns and antecedents agree.
- Verb tense is consistent.
- Modifiers are used appropriately.
- Word choice and sentence structure follow guidelines for localization.
- Long sentences are divided for readability and localization.

Punctuation, Capitalization, and Spelling

- Punctuation follows editorial and documentation set guidelines.
- Capitalization follows editorial and documentation set guidelines.
- Spelling follows editorial and documentation set guidelines.

Mechanics

- Typeface conventions are followed in all document elements.
- Product names are used correctly and consistently.
- Trademarks are used correctly and include appropriate attributions.
- New terms are defined and appear in a glossary if one is included.
- Abbreviations and acronyms follow editorial and localization guidelines.
- Numbers and symbols follow editorial and localization guidelines.
- Cross-references are punctuated correctly and refer to the intended target.
- Bulleted lists are used appropriately and follow editorial guidelines.
- Numbered lists and steps are used appropriately and are numbered correctly.
- Figures, tables, and examples are referred to in preceding text.
- Table continuations are noted correctly.
- Notes, Cautions, and Tips are used correctly.
- Footnotes are used correctly.
- Running footers and page numbers are correct.

Formatting and Layout

- Document conforms to house publications standards.
- Standard templates and formats are used.
- Page breaks and line breaks are effective.
- Page numbering scheme is appropriate for the type of book.

Illustrations

- Illustrations are consistent and sized appropriately throughout the book.
- Illustrations follow artwork and localization guidelines.
- Figure callouts are capitalized correctly and are in the correct font.

Proofreading

Proofreading is the last step that writers and editors can take to ensure further quality. Proofreading involves one final scan of the document for errors that might have been overlooked in previous reviews. The writer also might have introduced errors when incorporating new technical material or editorial comments. The proofreader's primary responsibility is to make sure that typographical errors, incorrect font usage, and formatting mistakes have not crept into the document.

Proofreading Guidelines

Front Matter

- Title page shows correct title and company name and address.
- Legal notice is current and trademarks, including third-party ones, are listed.
- Table of contents includes correct headings and page number references, and is formatted correctly.
- Figures, tables, and examples are listed in the front matter.
- The preface uses the correct template and contains correct chapter numbers, descriptions, and any required product-specific information.

Back Matter

- Appendixes are in the correct order.
- Templates and formats are used correctly in appendixes and glossaries.
- Glossary is correctly presented.
- Bibliography is correctly presented.
- Index is formatted correctly and contains no errors.

Grammar

- Sentences are complete.
- Subjects and verbs agree, and pronouns and antecedents agree.
- Verb tense is consistent.
- Modifiers are used appropriately.

Punctuation, Capitalization, and Spelling

- Punctuation follows editorial and documentation set guidelines.
- Capitalization follows editorial and documentation set guidelines.
- Spelling follows editorial and documentation set guidelines.

Mechanics

- Typeface conventions are followed in all document elements.
- Product names are used correctly and consistently.
- Trademarks are used and attributed correctly.
- New terms are italicized, defined, and appear in a glossary, if one is included.
- Abbreviations and acronyms follow editorial and localization guidelines.
- Numbers and symbols follow editorial and localization guidelines.
- Cross-references are punctuated correctly and refer to the intended target.
- Bulleted lists are used appropriately and follow editorial guidelines.
- Numbered lists and steps are used appropriately and are numbered correctly.
- Figures, tables, and examples are numbered correctly.
- Table continuations are noted correctly.
- Footnotes are used correctly.
- Page footers and numbers are correct.
- Change bars do not appear.

Formatting and Layout

- Document conforms to house publications standards.
- Standard templates and formats are used.

Illustrations

- Figure callouts are capitalized correctly and are in the correct font.
- Artwork is aligned correctly on the pages.

Planning Ahead for Editing

Writers need to allocate time for editing when creating a documentation plan. The nature of the document and the schedule determine how much editing is possible.

Consider these points:

- You can involve the editor as early as the research stage.

  The editor can help you with research on how similar products are handled and on who the audience is.
- The editor can help you prepare your documentation plan.

  Consult the editor if you want advice on overall organization. Go over your editing needs with the editor and include editing cycles in the schedule.
- The alpha review is a good time for a full developmental edit.

  Beta review is usually too late to make the kinds of changes that might come out of a developmental edit.
- A copy edit at the beta review can clean up grammar, spelling, and conformance to your company's style standards.
- Proofreading before the final release provides one last check for formatting issues and typographical errors.

If you are writing a white paper or other nonstandard document, plan to allow time for developmental edits and copy edits.

Submitting a Document for Editing

Before submitting a document for editing, complete these tasks:

- Run your document through a spelling checker.
- Run your document through a lint program if such a program exists for your authoring environment.
- Check cross-references.
- Include illustrations or indicate placement of illustrations.

Include a Request for Editing form with the document you submit. This form supplies information such as your name and phone extension, the stage of the document (alpha, beta, final release), and the name of the set to which the document belongs. See "Request for Editing Form" on page 413 for a sample form.

Editing Marks

Most editors are glad to explain any editing marks that were used in the edit. You can find an online guide to editing marks at the University of Colorado at Boulder web site. For an explanation of standard proofreaders' marks, see *The Chicago Manual of Style, Fourteenth Edition*, or *Merriam-Webster's Collegiate Dictionary*, which is available online at `http://www.m-w.com` (search for "proofreaders marks").

Creating a Style Sheet

Maintaining a *style sheet* can help you keep track of special spellings, terminology, punctuation, capitalization, and other document-specific words or formats.

A style sheet is where you and the editor can log the decisions made about product names, numbers, abbreviations and acronyms, hyphenation, and capitalization. If the document that you are writing or editing is part of a set, using a style sheet helps maintain consistency among the various documents.

When you create a style sheet, remember to pass it on to others who might benefit from it. These people might include writers of related documents, editors, illustrators, and production specialists.

A sample style sheet follows.

Editorial Style Sheet

Document title:

Project:

Writer:

Editor:

Date:

A B C	D E F	G H I
J K L	**M N O**	**P Q R**
S T U	**V W X**	**Y Z Numbers**

Cover capitalization, spelling, hyphenation.

(n) noun	(v) verb	(a) adj preceding noun
(pa) predicate adjective	(col) collective noun	(s) singular
(pl) plural	(TM) trademark	(R) registered trademark

Abbreviations

Trademarked Terms

Special Font Conventions

Miscellaneous Notes

◆ ◆ ◆ CHAPTER 12

Working With Illustrations

A good illustration transmits dense and complex information at a glance. It also helps a reader to retain more information. Combining text and images focuses attention and helps a reader filter data quickly for specific information. As documentation moves online, you might find yourself in the role of a screenwriter or director as you blend text, still images, video, audio, and animation in online multimedia documents.

A high level of image quality is essential to visual communication in any medium. From the printed page, to paint on canvas, to dots on a computer screen, similar rules apply. Even though the elements of good visual art can sometimes defy simple explanation, most people can say they know quality when they see it. In basic terms, the human eye should be able to scan a quality image without backtracking to retrace a path or disengaging to refocus. Quality images also contain balanced elements, consistent line weights, and a discernible flow.

Poor quality images are usually easy to identify, even though beauty is sometimes in the eye of the beholder. In general, poor quality images contain unbalanced proportions, jarring colors, jagged or crossed lines, or illegible or blurry elements. If the images in your document appear substandard, contact your illustrator or production specialist immediately for assistance.

This chapter discusses the following topics:

Working With an Illustrator

If your document contains new or complex visual elements, include an illustrator on your team. Illustrators are experts in visual communication and have access to specialized graphics tools, computer-aided design (CAD) models, and graphics repositories.

Involving an illustrator means that any work that is created can be shared with other writers. The work might also be added to an image library or catalog and be made to adhere to approved graphics standards.

The illustrator can do the following:

- Review your documentation plans and devise a visual strategy.
- Collaborate with you as the document changes throughout its life cycle.
- Resolve file format, platform, and tools issues.
- Maintain a consistent look and feel throughout the document.
- Reuse illustrations in other documents while maintaining quality.

Improving Quality and Accelerating Workflow

Writers who have illustration skills and tools should still work with an illustrator for several reasons.

- Time spent on complex drawings can compromise deadlines.

  Illustrators can usually handle difficult or complex drawings more efficiently and effectively.
- Artwork created by illustrators can be archived and shared with others.
- Visual consistency is enhanced when all work is done by illustrators who are equipped with the same tools and templates.
- Maintaining open relationships and communication with illustrators speeds up the production process and improves quality.

Contacting an Illustrator

Involve illustrators at the outset of a documentation project in the following instances:

- When you are unable to locate existing illustrations to meet your needs
- When existing illustrations must be modified
- When complex illustrations combining text layers and backgrounds must be created
- When you have concerns about the quality of illustrations in your document
- When you have a concept or a rough sketch that needs professional execution

Submitting an Illustration Request

When requesting help with illustrations, remember to include information that the illustrator on your team needs to know:

- Number of documents in your project and their identification numbers
- Estimated number of illustrations
- Marked hard copies or changes to existing illustrations
- Original source files for illustrations to be modified
- Hand-drawn sketches or concept diagrams
- Estimated production dates or deadlines

Understanding Illustration Files

Understanding illustration file types can help you choose the best approach for your project. Two main file types are used in illustrations: vector and raster.

Vector Images

Vector-based images contain mathematical image data based on x and y coordinates. Vector graphics can therefore be scaled to any size without quality loss and with a lower relative file size. Vector graphics are ideal for print use because they can be output to printers at any resolution setting. The most common vector-based file type extension is Encapsulated PostScript (`.eps`). Vector-based images are used for line art and composite images with typography. They must be *rasterized* to be used on the web.

Raster Images

Raster images are also known as *bitmaps*. Raster images contain a grid of colored pixels that are fixed in place. Unlike vector graphics, raster images cannot be easily scaled without distortion. Raster images are most useful for screen captures, digital photography, and web graphics. Common file extensions are `.tif`, `.gif`, and `.jpg`. Contact your illustrator or production specialist for help in understanding and choosing the best file format for your project.

For guidelines on rasterized screen captures, see "Creating Quality Screen Captures" on page 243.

Types of Illustrations

Several different types of illustrations are commonly used by technical communicators. The following illustration types are described in the following sections:

- "Diagrams" on page 232
- "Line Art" on page 232
- "Screen Captures" on page 233
- "Photographs" on page 234
- "Animations" on page 235

Diagrams

Diagrams encompass a wide range of illustration uses, from simple flow diagrams and presentation aids to complex technical diagrams.

Line Art

Line art illustrations of hardware are usually based on engineering diagrams or 3-D CAD drawings. When possible, try to use the original vector-based source file. The vector format preserves quality and also provides the most versatile image editing options.

EXAMPLE 12–1 Line Art

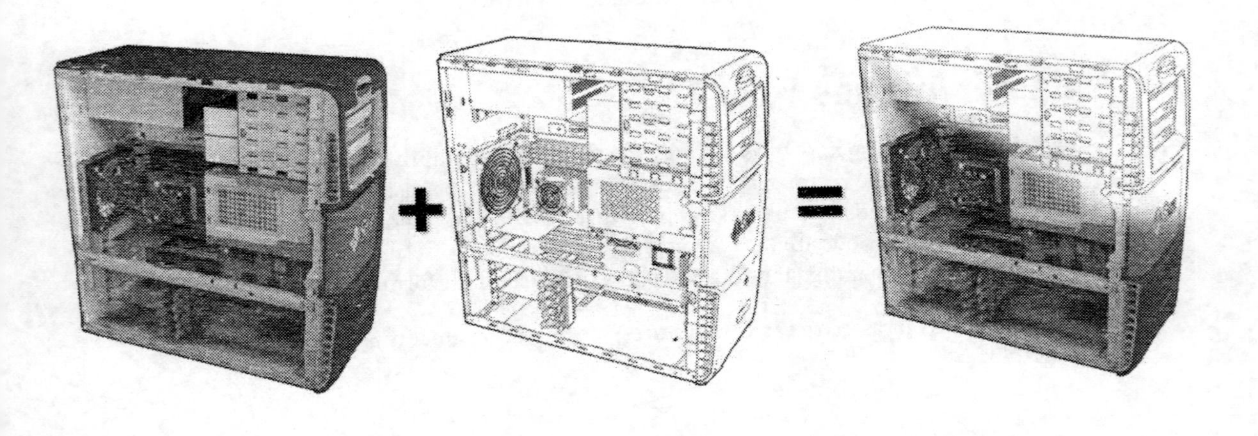

Screen Captures

A screen capture is always created as a raster image. Screen captures represent an exact snapshot of the dots that are displayed on a computer monitor. Once created, screen captures cannot be easily resized without distortion. A technical illustrator can advise you on the proper way to scale and crop screen captures, if necessary.

See "Creating Quality Screen Captures" on page 243 for more information.

EXAMPLE 12–2 Screen Capture

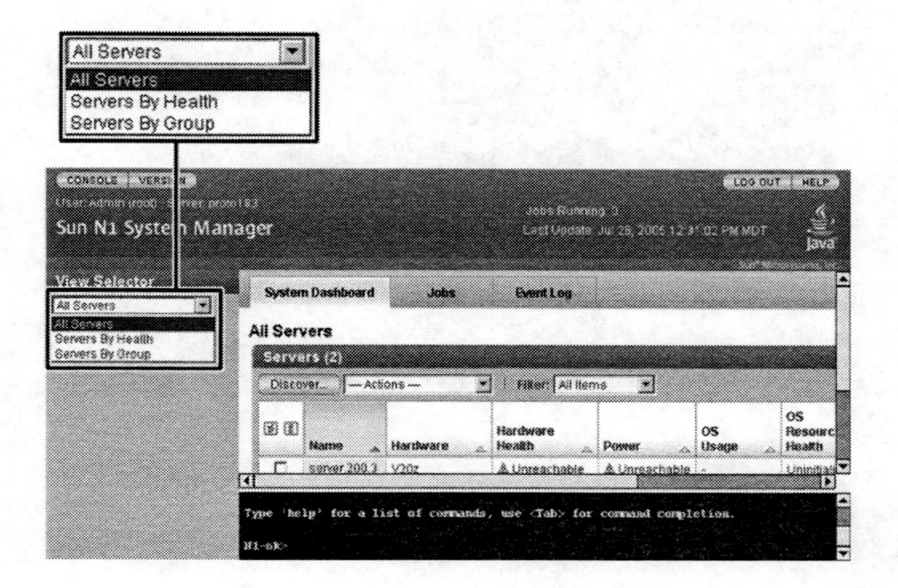

Photographs

Photographs can be captured in raster format from a digital camera or scanned from a photo print. File sizes can be large, depending on the resolution, size, and number of colors used. A technical illustrator can edit and manipulate photographs.

EXAMPLE 12–3 Photograph

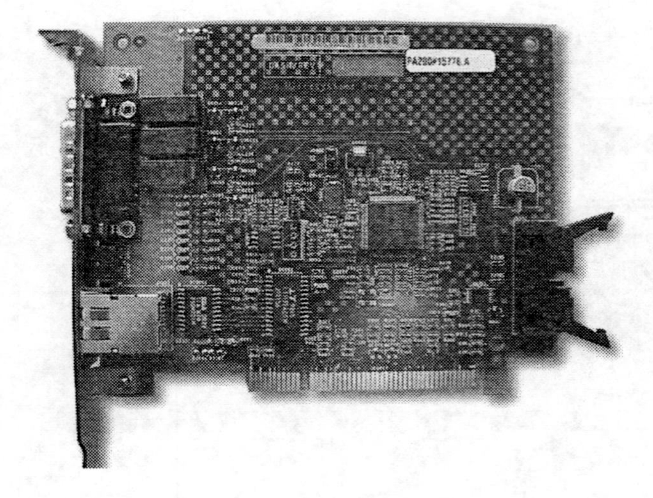

Animations

Animations can be created as vector graphics or raster images. The file format and tools that illustrators use to create animations are often tailored to the delivery requirements of the end user. Animations can affect download times, so they must be well designed and used sparingly. An illustrator's help is critical to the design of effective and compact animations.

Animations, media elements, and presentation graphics often require the use of text elements that are integral to the project's overall design. The guidelines for technical callouts might not apply in many of these cases, but should still be referred to as general guidelines for text use.

EXAMPLE 12–4 Animation

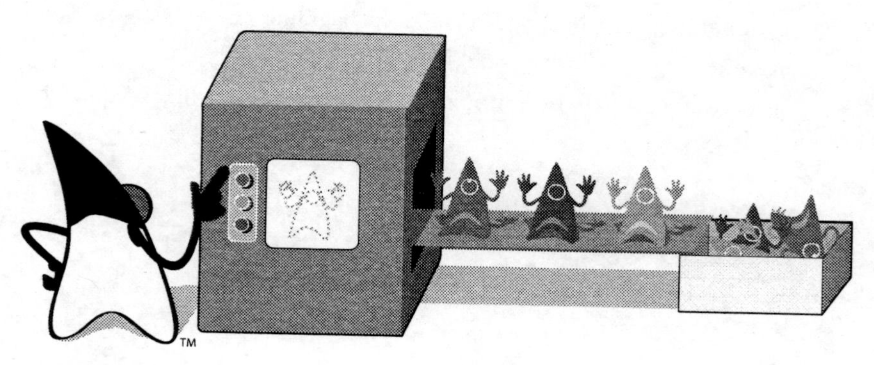

Placing Illustrations

Your illustrations must appear to be an integral part of your document flow. You can achieve this effect in print and on the web by using consistent placement, spacing, and alignment. Some publication tools handle image placement automatically but others do not.

This section contains guidelines for spacing your illustrations and for aligning illustrations with the other elements in your document.

Placement in Relation to Sentences

Follow these guidelines when placing illustrations in your document:

- Use a complete sentence ending in a period when introducing an illustration, not a phrase ending in a colon.
- Refer to the illustration's position on the page in the document flow, for example, "The following illustration shows a diagram of the host server hierarchy."In subsequent references to the same illustration, use "preceding" or "following" as appropriate.

- Refer to the illustration's number only if it is not close to the introductory text.
- Do not insert an illustration between the beginning and the end of a sentence.

  Incorrect:

  This diagram

  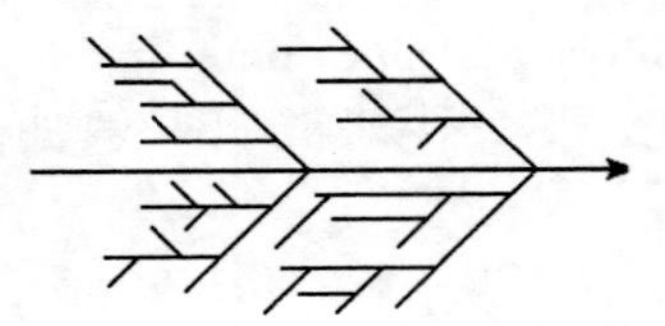

  is commonly called a fishbone or Ishikawa diagram.

  Correct:

  The following diagram is commonly called a fishbone or Ishikawa diagram.

  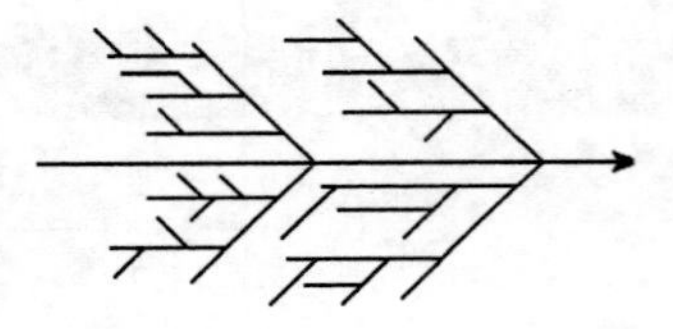

Spacing in Print

The following table contains guidelines for how much spacing to allow between an illustration and printed text above and below it.

TABLE 12–1 Illustration Spacing Guidelines

Element	Minimum Distance	Maximum Distance
Text above	0.25 inch (0.64 cm)	0.75 inch (1.9 cm)
Text below	0.25 inch (0.64 cm)	0.75 inch (1.9 cm)

Alignment in Print

Print production tools often position illustrations into the text flow automatically. If not, align illustrations flush-left to the margin of the text column in most instances. If the drawing is wider than the text column, align the drawing flush-left to the page margin.

Online Alignment and Spacing

When placing illustrations in online documents, apply the following guidelines:

- Follow a consistent spacing and alignment method on all of your web pages to create a balanced look and feel.
- Follow any preexisting guidelines for your project that might already be documented by previous webmasters or content creators in your group.
- Know the minimum sizing and performance limitations of your audience's web browser. Keep your design within those parameters.

Writing Captions for Illustrations

Captions help set the context of the illustrations within the document. Captions also uniquely identify each illustration for cross-referencing. This section contains guidelines for writing unique captions.

When to Use Captions

Use captions when doing so can help your reader locate the illustration as a general reference element. For example, an illustration of a common menu might not need a caption, whereas a conceptual illustration would require a caption.

If the illustration is used in the context of a step and is not for general reference, do not use a caption. An example of introductory text for such an illustration might be, "The following figure shows the Network Status window after you have completed this step." Illustrations for general reference should appear in conceptual text rather than procedural steps.

Use captions in the following instances:

- When you want to cross-reference the illustrations
- When you want to generate a list of figures

Guidelines for Writing Captions

Follow these guidelines when writing captions:

- Use the same capitalization style for figure captions as you use for section headings.
- Do not start a figure caption with an article. For example, write "File Menu," not "The File Menu."
- Limit the figure caption text to one line.

- Introduce the context of the illustration to the reader in the text preceding the illustration.
- Ensure that the figure captions are numbered sequentially throughout the document if your authoring tool does not number figure captions automatically.

Writing Callouts for Illustrations

Callouts are text elements that describe the details of a technical illustration. They help connect illustrations to the surrounding text. This section contains guidelines for integrating callouts and illustrations.

When creating callouts, follow these guidelines:

- Use the minimum number of callouts necessary.
- Use the shortest and fewest words possible.
- Move lengthy callouts to a note or legend outside the drawing.
- Omit punctuation except for a question mark when needed.
- Capitalize the words in a callout as you would the words in a sentence.

  Conversely, capitalize textual elements intended to serve as headings, or labels, for conceptual elements in illustrations as you would the words in a heading.
- If your document will be translated, leave ample space vertically and horizontally to facilitate future translation.
- Capitalize proper nouns, abbreviations, and acronyms as you do in text.

The following figure shows examples of capitalization in callouts.

EXAMPLE 12–5 Callout Capitalization

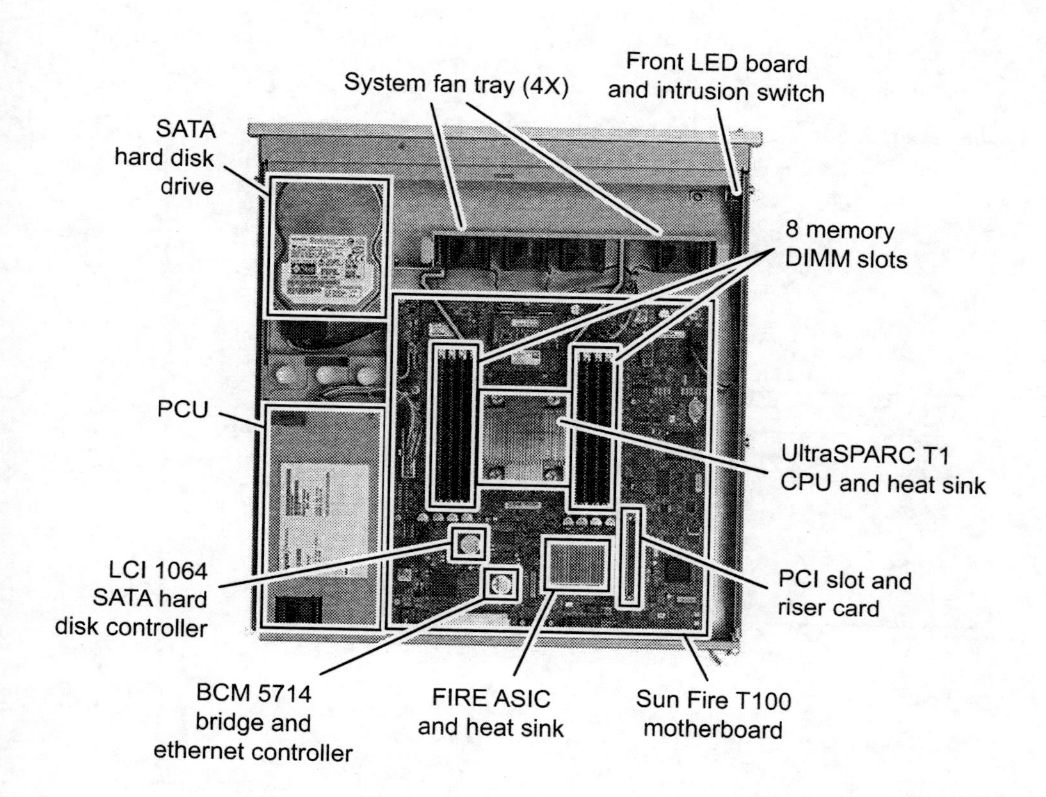

Callout Style

The following table contains style specifications for callouts.

TABLE 12–2 Callout Specifications

Specification	Description
Typeface	Helvetica, or Courier for monospace
Type weight	Regular (not bold)
Alignment	Left, right, or centered, based on relationship to graphic element
	In diagrams, including flow diagrams, callouts are centered
Numbering	Clockwise, from a logical starting position

Placement of Callouts

Arrange callouts consistently throughout a document. Place callouts far enough apart to be read clearly. Space callouts evenly so that the illustration does not look cluttered and is not difficult to read.

The following examples show callouts with appropriate spacing.

EXAMPLE 12–6 Callout Spacing

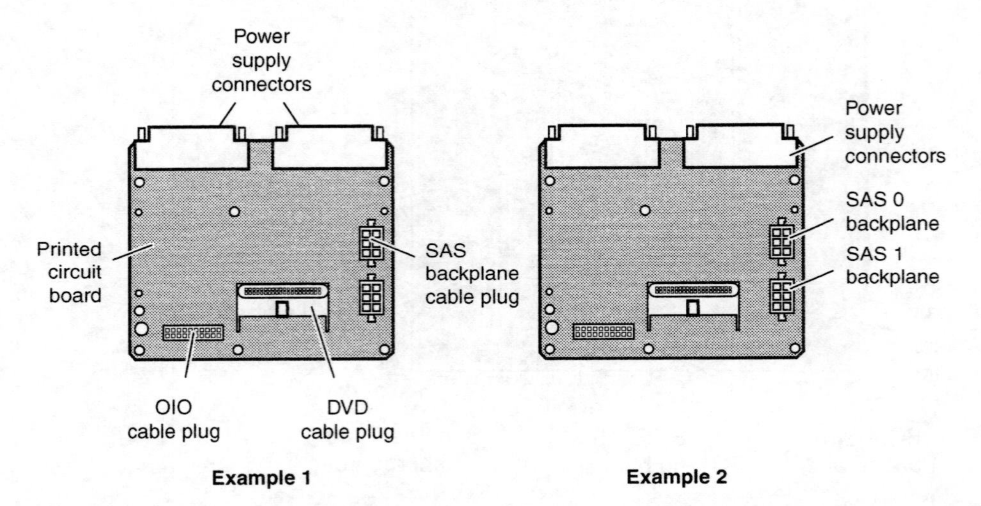

When faced with multiple callouts in a drawing, or when space is limited, consider other textual alternatives. If a table would convey the information equally well, use a table instead of an illustration. Tables are simpler to create and maintain.

Similar to text, callouts must be translated when a document is localized. See "Create Callouts That Are Easy to Translate" on page 179 for more information about making sure that your callouts can be easily translated. Text elements may expand up to 30 percent when translated into another language, so callouts should be planned accordingly.

Another option involves creating a legend in the lower part of the illustration. A legend usually references numbered callouts within the illustration, or uses standard symbols as a key. The following figure shows numbered callouts in a legend.

EXAMPLE 12–7 Numbered Callouts in a Legend

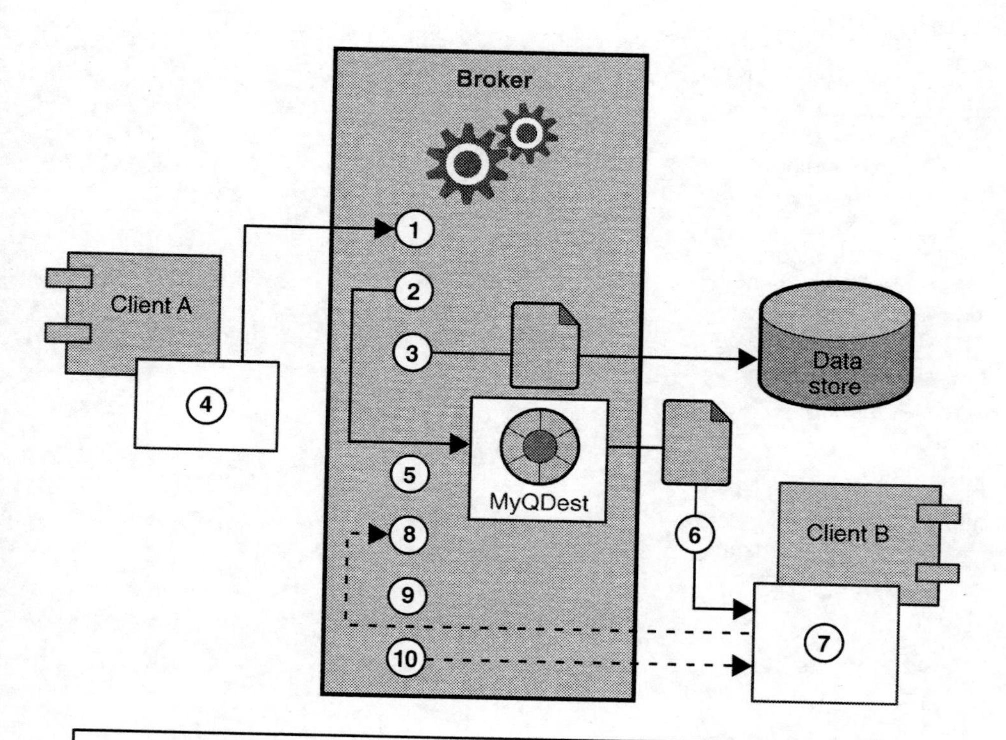

1. Input to Broker
2. Message generated
3. Message output to Data store
4. Client A runtime
5. Route to MyQDest
6. Output to runtime
7. Client B runtime
8. Message to Broker
9. Message to Broker
10. To Client B

The following figure shows callout placement in a flow diagram.

EXAMPLE 12–8 Callouts in a Flow Diagram

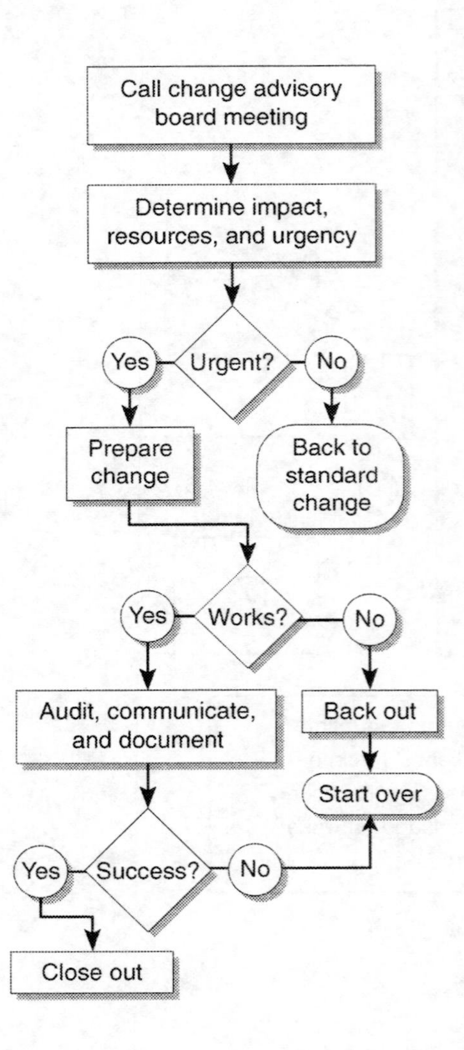

Creating Quality Screen Captures

Including screen captures in your documents is an effective way to help communicate graphical user interface (GUI) operations.

Guidelines for Creating Screen Captures

When creating screen captures, follow these guidelines:

- Check your capture settings.

  Screen capture tools often default to 72 dots per inch (dpi) when capturing images. Make sure that you capture your images at an acceptable size and resolution for your target document.

  Note that the dpi of printed raster images affects physical size. For example, a 72-dpi capture shrinks 50 percent when printed at 144 dpi. Print standards often require 300 dpi or higher, while 72 dpi is sufficient for online viewing.
- When creating screen captures for print, use a printer-friendly color palette or a greyscale desktop color scheme.
- Save your screen capture in a high-quality format such as `.tif` (for print) or `.gif` (for the web) whenever possible.
- Do not resize the screen capture by dragging or stretching its boundaries, which can cause image distortion.

Using Screen Captures as Guideposts Only

When including screen captures in online documents, place them at chosen points to serve as guideposts in navigation. These guideposts can help readers remain properly oriented and confirm that the document is proceeding as expected.

Including too many screen captures can be redundant. Avoid using screen captures in the following instances:

- When forms and dialog boxes are designed consistently across the GUI
- When GUI forms are self-explanatory and require no further explanation
- When data entry fields are labeled clearly and consistently
- When the GUI provides adequate feedback when used incorrectly

  For example, when users enter invalid data, feedback is generated.

Ensure that screen captures in online documents are not mistaken for actual software interfaces. Even sophisticated readers can confuse the two. For guidance on cropping, highlighting with color, and other strategies to make the distinction more obvious, contact your illustrator or production expert.

Creating Leader Lines

A *leader* points from a callout to a specific part of an illustration. This section contains instructions for creating leader lines and guidelines for using leader lines in illustrations.

Leader Style

The following table contains recommended specifications for leader lines.

TABLE 12–3 Leader Line Specifications

Specification	Description
Line weight	0.4 point
Length	Minimum length: 0.25 inch (0.64 cm) Maximum length: as short as possible
Angle	Consistent in drawing when possible Never perpendicular or parallel to text Avoid blending with illustrations

Additional Guidelines

When creating leader lines, follow these guidelines:

- Arrange callouts so that leader lines cross as few illustration lines as possible.
- Arrange leader lines so that the lines never cross.
- Terminate the leader line on the element.
- Begin the leader line at the corner closest to the callout.

Simplifying Online Illustrations

Complicated illustrations that work effectively in print work less well online because of screen readability problems. What is legible in print might not display clearly online.

To keep illustrations simple and to facilitate online reading, apply the following guidelines:

- Increase line weights and text size.

  Fine lines and small text are difficult to read online.
- Use fewer details online.

  Due to low resolution on some monitors, details that are effective in printed illustrations might be unreadable online.
- Reduce the size of large illustrations and crop them tightly.
- Try to use tables instead of illustrations.

  If a table would convey the information equally well, use a table instead of an illustration. Tables are simpler to create and maintain.

◆ ◆ ◆ CHAPTER 13

Writing Alternative Text for Nontext Elements

The Section 508 Amendment to the U.S. Rehabilitation Act requires that electronic and information technology that is developed or purchased by the federal government be accessible by people with disabilities. This chapter describes how to create documents that meet one of the key Section 508 accessibility requirements. This key requirement is to create text equivalents, referred to as *alternative text*, for each nontext element in the document. This chapter explains how to write alternative text.

This chapter discusses the following topics:

Section 508 Requirements Overview

United States government purchases of information technology are required by federal law to be accessible by people with disabilities. Examples of disabilities include vision impairment, hearing impairment, and so on. Requirements focus on the need for assistive technologies that enable people with disabilities to use software products and to get the full benefit from documentation.

Assistive Technologies

To meet Section 508 accessibility requirements, products and documentation must provide support for assistive technology tools that enable specialized input and output capabilities.

Examples of assistive technology tools include the following:

- **Screen reader.** This tool enables vision-impaired users to navigate through applications, determine the state of controls, and "read" text by using text-to-speech conversion.
- **Screen magnifier.** This tool enables users to enlarge the screen display.
- **Voice command and control.** This tool enables users to control the system by using spoken commands.

Guidelines for Meeting Section 508 Requirements

To ensure that documents meet Section 508 accessibility requirements, provide alternative text to describe nontext elements. Follow these basic guidelines:

- Identify the nontext elements that are included in a document.

  Nontext elements include graphics, mathematical equations, tables, and multimedia content.
- Create short alternative text for each nontext element.

  Short alternative text, which *cannot* exceed 150 characters, is required for each nontext element. This text enables assistive access tools to "read" the nontext element.
- Create long alternative text for select nontext elements, only if necessary.

  Sometimes, a nontext element cannot be adequately described in fewer than 150 characters. In such cases, long alternative text, which can exceed 150 characters, can be used in conjunction with short alternative text. Long alternative text *must* be accompanied by short alternative text.

General Guidelines for Writing Alternative Text

Document accessibility is addressed by providing alternative text descriptions for nontext elements. When writing alternative text, do the following:

- Describe callouts and other relevant information in a logical order.
- Use the same style guidelines that you use for the rest of the document.
- Do not add HTML tags, character styles, or markup of any kind to alternative text.

  Specific formatting is not interpreted in alternative text, so such markup appears as is.
- Identify the graphic type prominently in the alternative text, namely at the beginning or at the end, but not in the middle.
- If the *only* occurrence of a trademarked term appears in the alternative text, append the appropriate trademark symbol in the format (R) or (TM) to the term.

  Do not mark the occurrence in the alternative text if the trademarked term is used elsewhere in the body of the document.

The first step in providing alternative text is to determine the context of the nontext element.

Determining the Context of a Nontext Element

The existence and amount of context that a nontext element has in the surrounding text can help you determine the following:

- How much alternative text to write
- What kind of alternative text to write

Look at each nontext element and the context that surrounds it. The context usually appears within the same section as the element. The three main context levels are defined as follows:

- **Existing context.** A description of the nontext element or its purpose appears in the body of the document.

  Provide short alternative text for this context level, as described in "Writing Short Alternative Text" on page 247.
- **Limited existing context.** A limited description of the nontext element or its purpose appears in the body of the document.

  Provide short alternative text for this context level, as described in "Writing Short Alternative Text" on page 247.
- **No existing context.** No description of the nontext element appears in the body of the document and the purpose of the nontext element is unclear.

  Determine whether the nontext element serves a useful purpose in the document. If the nontext element serves no useful purpose, remove the element from the document.

  If the element serves a useful purpose, provide short alternative text as described in "Writing Short Alternative Text" on page 247. If you consider long alternative text to be necessary, see "Writing Long Alternative Text" on page 248.

Note – If your document is early enough in the production cycle, consider adding contextual text for the nontext element in the body of the document. The description will then be available to regular readers as well as those readers who are accessing alternative text.

Writing Short Alternative Text

You must write short alternative text for every nontext element. Such text is the value of the `ALT` attribute of the `<IMG>` tag. Conversion tools for several authoring environments make this change for you when converting from other formats to HTML. Short alternative text cannot exceed 150 characters.

This text provides a brief description of one of the following:

- What the nontext element represents
- What the nontext element looks like
- What purpose the nontext element serves in the document

Avoid duplicate information, irrelevant details, or inconsistent terminology.

Writing Long Alternative Text

Note – Long alternative text is not supported by all authoring tools and browsers. Even when supported, long alternative text is available only to users who use screen readers.

You can optionally use long alternative text to provide very detailed descriptions of nontext elements. Complex graphics, diagrams, and flow diagrams might be good candidates for long alternative text. The long alternative text is placed in a separate file. The name of this file is the value of the `LONGDESC` attribute of the `<IMG>` tag.

The alternative text examples that appear in "Complex Graphics" on page 250 and in "Diagrams" on page 254 include examples of long alternative text.

You write long alternative text just as you write short alternative text, but long alternative text can exceed the 150-character limit.

Writing About Nontext Elements

You use the same writing guidelines to describe most nontext elements that you do to describe text. This section includes writing guidelines for the following graphic types:

- "Simple Graphics" on page 249
- "Complex Graphics" on page 250
- "Diagrams" on page 254
- "Mathematical Equations" on page 259
- "Multimedia Content" on page 260

Tip – Avoid terms such as "annotation," "blowup," and "exploded view" in alternative text. Such terms rarely provide information relevant to the alternative text.

The examples in the following subsections show graphics in isolation along with alternative text. The descriptions of the type of context identified in the Notes column of the example tables use these terms:

- **Existing context assumed.** The nontext element might clarify the information presented or might add visual interest to a document. Alternative text requirements can be satisfied with minimal text.
- **Limited existing context assumed.** The nontext element might clarify the information presented. Provide more detail about the content of the nontext element if the details are not available in the existing context. Alternative text requirements might be minimal or somewhat more.

- **No existing context assumed.** The nontext element is vital to the understanding of the information presented, but it is not described by existing context. Alternative text requirements can probably only be satisfied by extensive text.

Note – If your document is early enough in the production cycle, consider adding contextual text for the nontext element in the body of the document. The description will then be available to regular readers as well as those readers who are accessing alternative text.

Most of the examples show alternative text for all three contexts. These variations are numbered, and character counts for each version are provided in the adjacent Notes column.

Simple Graphics

Simple graphics usually represent single entities, such as symbols, logos, icons, and figures. *Symbols* are graphics that represent functions or operations. *Logos* are graphics that identify a company or product. *Icons* are graphics that represent applications or functions within an application. Symbols, logos, and icons are not commonly accompanied by captions. They might be separated from the document text or placed within a sentence as an inline graphic.

Symbols and icons might be *static* or *dynamic*. A static symbol or static icon is used only as an illustration. A dynamic symbol or dynamic icon might be used as a button or link to perform another task. For example, a dynamic symbol might be used to launch a multimedia application.

When writing alternative text for simple graphics, follow these guidelines:

- Identify the graphic as a symbol, logo, icon, or figure.
- Describe the purpose of the graphic.

  For example, explain an icon's state as active or inactive.
- Provide only a name for common symbols, logos, and icons.

  Provide more detail for uncommon graphic elements.

Note – Provide alternative text for bullets and glyphs because they are graphics. Set a null description (`ALT=""`) to signify to assistive technologies that the bullets and other glyphs are unimportant elements.

EXAMPLE 13–1 Simple Graphics Text Equivalents

Simple Graphics	Notes
SCSI symbol.	Short alternative text: 12 characters.

EXAMPLE 13–1 Simple Graphics Text Equivalents *(Continued)*

Simple Graphics	Notes
Sun Creator3D system logo for Ultra(TM) 1 workstations.	Short alternative text: 52 characters.
 Netscape icon.	Short alternative text: 14 characters.
 Figure showing Start, Stop, and Reset buttons on the Main window.	Short alternative text: 64 characters.

Complex Graphics

Complex graphics present more complex ideas by means of screen captures and technical illustrations. Such graphics are sometimes accompanied by a figure caption. Complex graphics can be cropped, contain callouts, be annotated, show action, be exploded views, or show cut-away views.

When writing alternative text for complex graphics, follow these guidelines:

- Identify the graphic as a screen capture or figure.
- Include information about callouts, if applicable.
- Describe the purpose of the graphic.

  For example, explain the fields displayed.

- Describe the information in a logical order, if possible.

  For example, describe elements from top to bottom and left to right.

The following examples provide samples of accessibility text for complex graphics. The sample nontext element is followed by a table. The first column in the table shows the associated alternative text for the nontext element. The second column provides notes about the alternative text.

EXAMPLE 13–2 Screen Capture With No Figure Caption

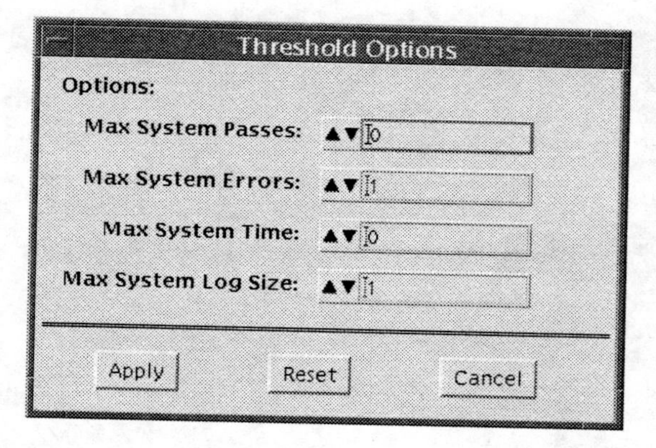

Alternative Text	Notes
(1) Screen capture of the Threshold Options dialog box.	(1) Existing context assumed. Short alternative text: 51 characters.
(2) Screen capture of Threshold Options dialog box showing Max System Passes, Errors, Time, and Log Size options. Buttons are Apply, Reset, and Cancel.	(2) Limited existing context assumed. The short alternative text does not mention the default values of the options because, in this case, that information is described in the text. Short alternative text: 147 characters.
(3a) Screen capture of the Threshold Options dialog box. (3b) The Threshold Options dialog box has four options and three buttons. The first option is Max System Passes, which has a default value of 0. The Max System Errors option has a default value of 1. The Max System Time option has a default value of 0. The last and fourth option, Max System Log Size, has a default value of 1. The buttons are Apply, Reset, and Cancel.	(3) No existing context assumed. Could not list all of the options and buttons without exceeding the 150-character limit, so long alternative text was added. (a) Short alternative text: 51 characters. (b) Long alternative text: 365 characters.

Tip – Some screen readers read the figure caption before they read the alternative text, while others read the caption after the alternative text. Be aware of how the alternative text would sound if read before and after the caption to avoid confusing, unintentionally funny, or embarrassing word combinations.

EXAMPLE 13–3 Composite Screen Capture With Figure Caption

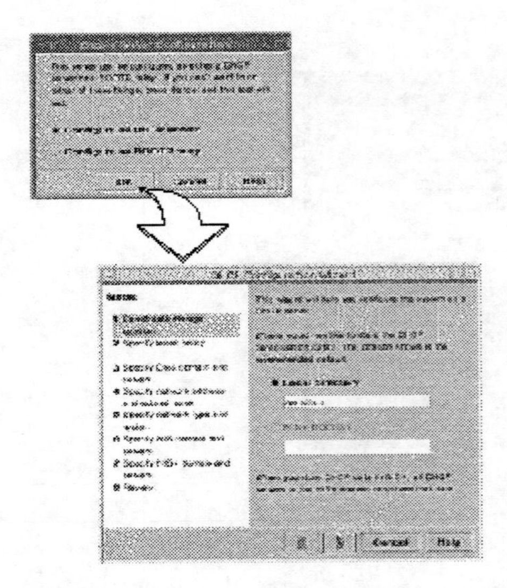

FIGURE 2–3 Opening the DHCP Configuration Wizard

Alternative Text	Notes
(1) Screen capture showing the Choose Server Configuration dialog box opening the DHCP Configuration Wizard.	(1) Existing context assumed. Short alternative text: 104 characters.
(2) Screen capture showing the Choose Server Configuration window, with Configure as DHCP Server and OK buttons selected, opening the DHCP Config Wizard.	(2) Limited existing context assumed. Short alternative text: 149 characters. To keep the number of characters below 150, "Configuration dialog box" has been replaced with the term "Configuration window." In addition, "DHCP Configuration Wizard" has been replaced with the term "DHCP Config Wizard."

EXAMPLE 13–3 Composite Screen Capture With Figure Caption *(Continued)*

Alternative Text	Notes
(3a) Screen capture showing the Choose Server Configuration dialog box opening the DHCP Configuration Wizard. (3b) Selecting the Configure as DHCP Server button and then clicking the OK button on the Choose Server Configuration dialog box opens the DHCP Configuration Wizard dialog box.	(3) No existing context assumed. Could not describe the action without exceeding the 150-character limit, so long alternative text was added. (a) Short alternative text: 104 characters. (b) Long alternative text: 171 characters.

EXAMPLE 13–4 Annotated Technical Illustration

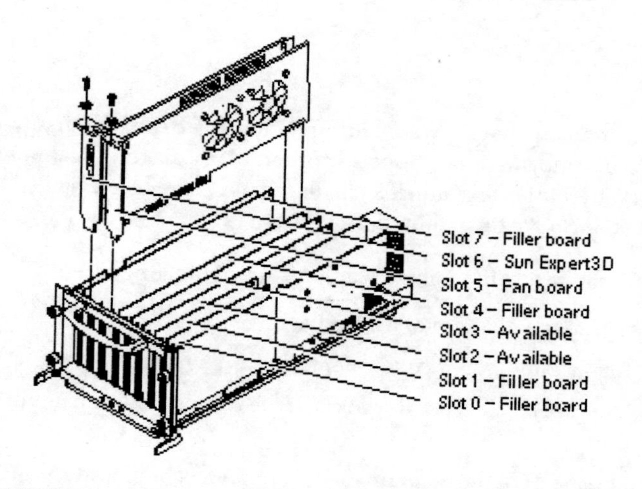

FIGURE 2-4 Slot Locations in a Sun Fire™ 6800 I/O Assembly

Alternative Text	Notes
(1) Figure showing Sun Fire(TM) 6800 I/O assembly slots and the boards that fill them.	(1) Existing context assumed. Short alternative text: 82 characters.
(2) Figure showing Sun Fire(TM) 6800 I/O assembly slots and the boards that fill them: Slot 6 has a Sun Expert3D card, and Slot 5 has a fan board.	(2) Limited existing context assumed. Short alternative text: 142 characters.

EXAMPLE 13–4 Annotated Technical Illustration (*Continued*)

Alternative Text	Notes
(3a) Figure showing Sun Fire 6800 I/O assembly slots and the boards that fill them. (3b) Figure showing which boards must fill the Sun Fire 6800 I/O assembly's seven slots: Slot 0 and Slot 1 are filled with filler boards. Slot 2 and Slot 3 are available for miscellaneous cards. Slot 4 has a filler board. Slot 5 has a fan board. Slot 6 contains the Sun Expert3D card. Slot 7 has another filler board.	(3) No existing context assumed. Could not describe the boards and slots without exceeding the 150-character limit, so long alternative text was added. (a) Short alternative text: 78 characters. (b) Long alternative text: 316 characters. In the long alternative text, the boards are listed in slot order from lowest to highest. The boards are also listed in a consistent manner, to make the information easier to follow for readers who are using assistive technology tools.

Diagrams

A *diagram* is a line drawing that shows the interrelationship of objects. A *flow diagram* is a diagram that presents information in a specific order. Many diagrams cannot be adequately described by short alternative text alone. Such diagrams can be accompanied by optional long alternative text *in addition to* the required short alternative text.

When writing alternative text for diagrams, follow these guidelines:

- Identify the graphic as a diagram or a flow diagram.
- Describe the purpose and major components of the diagram.

  If the components are labeled, use the labels. If the components are important but not labeled, describe them.
- Do not describe aspects of the diagram that are not relevant to understanding the diagram.

  For instance, the size or shape of a component in the diagram is usually irrelevant.

EXAMPLE 13–5 Standard Diagram

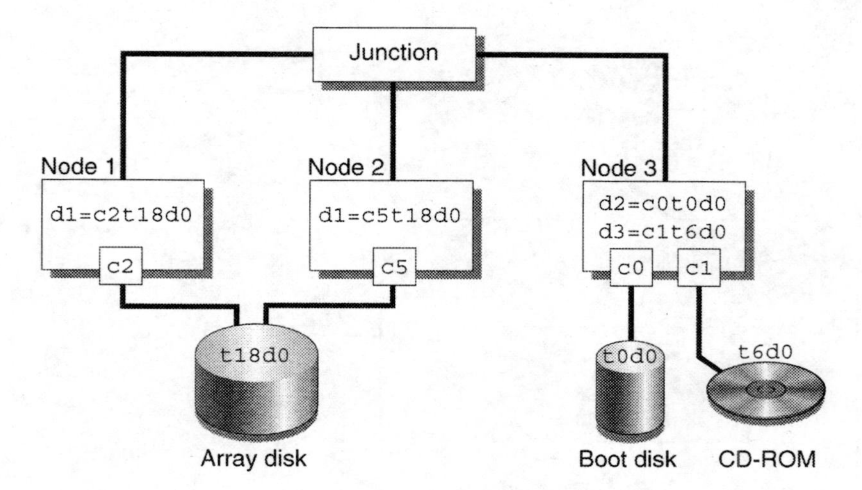

FIGURE 2-5 DID Numbering for a Three-Node Cluster

Alternative Text	Notes
(1) Diagram showing device ID numbering for a three-node cluster.	(1) Existing context assumed. Short alternative text: 61 characters.
(2) Diagram showing device ID numbering for a three-node cluster: Node 1 is d1=c2t18d0. Node 2 is d1=c5t18d0. Node 3 is d2=c0t0d0 and d3=c1t6d0.	(2) Limited existing context assumed. Short alternative text: 143 characters.
(3a) Diagram showing device ID numbering for a three-node cluster. (3b) Diagram showing device ID numbering for a three-node cluster. Array disk connects to Node 1 and Node 2. Both the boot disk and the CD-ROM connect to Node 3. Node 1, Node 2, and Node 3 all connect to the Junction. Array disk value is t18d0, boot disk value is t0d0, and CD-ROM value is t6d0. Node 1 value is d1=c2t18d0. Node 2 value is d1=c5t18d0. Node 3 values are d2=c0t0d0 and d3=c1t6d0.	(3) No existing context assumed. Could not describe the diagram without exceeding the 150-character limit, so long alternative text was added. (a) Short alternative text: 61 characters. (b) Long alternative text: 390 characters.

EXAMPLE 13–6 Complex Diagram

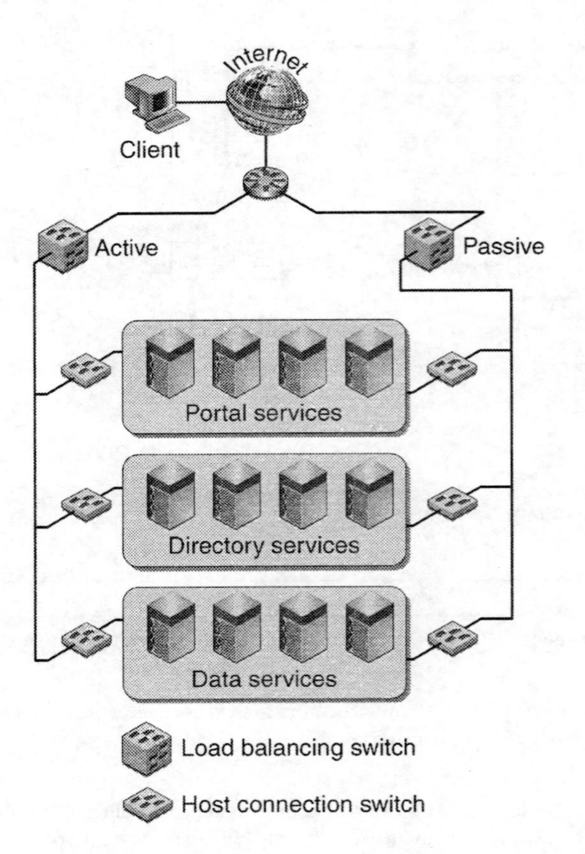

FIGURE 2-6 Service Delivery Network Components

Alternative Text	Notes
(1) Diagram showing components of the service delivery network.	(1) Existing context assumed. Short alternative text: 59 characters.
(2) Diagram showing active and passive load balancing switches connecting through host connection switches to portal, directory, and data services.	(2) Limited existing context assumed. Short alternative text: 143 characters.

EXAMPLE 13–6 Complex Diagram *(Continued)*

Alternative Text	Notes
(3a) Diagram showing components of the service delivery network. (3b) Host connection switches connect portal, directory, and data services to two load balancing switches. One load balancing switch is active. The other is passive. The load balancing switches, in turn, connect to a router that connects the entire service delivery system to the Internet. The client system connects directly to the Internet.	(3) No existing context assumed. Could not describe the details of the diagram without exceeding the 150-character limit, so long alternative text was added. (a) Short alternative text: 59 characters. (b) Long alternative text: 339 characters.

EXAMPLE 13–7 Flow Diagram

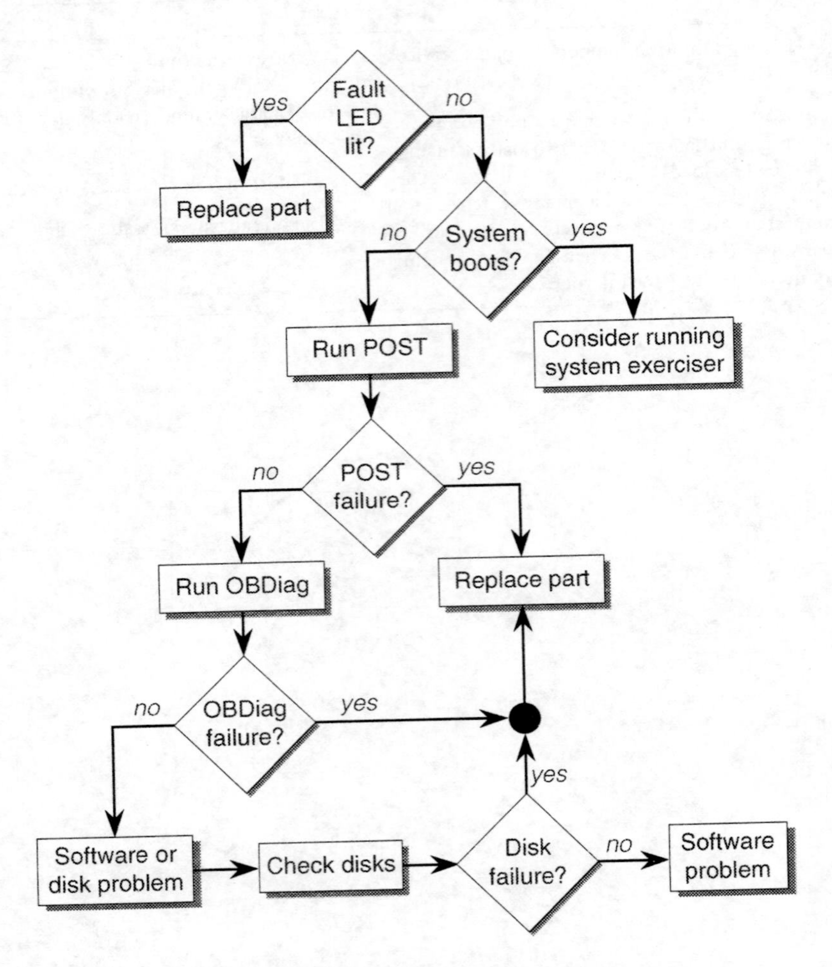

FIGURE 2-7 Choosing a Tool to Isolate Hardware Faults

Alternative Text	Notes
(1) Flow diagram showing the process of fault isolation.	(1) Existing context assumed. Short alternative text: 52 characters.

EXAMPLE 13–7 Flow Diagram *(Continued)*

Alternative Text	Notes
(2a) Flow diagram showing the process of fault isolation. (2b) Flow diagram showing fault isolation. If the Fault LED is lit, you must replace the part. If not, determine if the system boots. If the system boots, consider running system exerciser. If the system does not boot, run POST. If POST fails, replace the part. If POST succeeds, run OpenBoot(TM) Diagnostics (OBDiag). If OBDiag fails, replace the part. If OBDiag succeeds, you must check the disks for disk failure. If a disk failure occurred, replace the part. If no disk failure occurred, then software is the reason for the fault.	(2) No existing context assumed. Could not describe the flow diagram without exceeding the 150-character limit, so long alternative text was added. (a) Short alternative text: 52 characters. (b) Long alternative text: 530 characters. The shape of the figures in the flow diagram are omitted because decisions are dealt with in the text. The first occurrence of the trademarked term appears in alternative text, so it is marked with (TM).

Mathematical Equations

Many complex mathematical equations defy simple translation into words. To satisfy Section 508 requirements, describe the equation as a mathematical equation or as a graphic that shows a mathematical equation.

When writing alternative text for mathematical equations, follow these guidelines:

- Use the generic alternative text "Mathematical equation" or "Graphic showing mathematical equation" to describe the equation.
- Describe the purpose of the equation as if you were writing a caption title for the equation.

EXAMPLE 13–8 Mathematical Equations

Mathematical Equations	Notes
$$\text{allocation}_{\text{project}^i} = \frac{\text{shares}_{\text{project}^i}}{\sum_{j=1...n} (\text{shares}_{\text{project}^j})}$$ j is the index among all active projects Graphic showing mathematical equation.	Short alternative text: 39 characters.

EXAMPLE 13–8 Mathematical Equations *(Continued)*

Mathematical Equations	Notes
$\text{allocation}_{\text{project}_x^i} = \dfrac{\text{shares}_{\text{project}_x^i}}{\sum_{j=1\ldots n}^{\textit{processor set X}} (\text{shares}_{\text{project}^j})}$ j is the index among all active projects that run on processor set X Mathematical equation shows formula for how FSS scheduler calculates per-project allocation of CPU resources for projects running on processor sets.	Short alternative text: 149 characters.

Multimedia Content

Multimedia content presents information in a combination of media formats, which can include audio, video, simulation, animation, and text. Multimedia content usually supplements written material. For example, a remove-and-replace service procedure for a hardware component might also serve as the basis for a scripted video or animation of that procedure.

The first use of a multimedia icon in a document is usually as a static icon in the preface. This first use of the graphic introduces the concept of multimedia links that appear later in the text. See "Simple Graphics" on page 249 for examples of static icons.

Alternative text is not needed if the material is presented in the surrounding text. However, alternative text must still be provided for the multimedia icon.

When writing alternative text for multimedia content, follow these guidelines:

- Identify the graphic as a multimedia icon that has an underlying link.
- Describe the specific type of multimedia presented, for example, a movie or an audio file, and its purpose.
- Include the name of the movie or audio file.
- Provide a transcript of the audio, video, or animation content if that information is not already covered in another way in the document.

  For example, if closed captioning or audio descriptions are provided, a transcript is not required.

Example 13–9, Example 13–10, and Example 13–11 show examples of multimedia icons, animations, and simulations, respectively.

EXAMPLE 13–9 Multimedia Icons

Alternative Text	Notes
(1) Multimedia icon that can be selected to view a ShowMe(TM) How video demonstration of the "Removing the Power Supply" procedure.	(1) Existing context assumed. Gives the name of the movie. Short alternative text: 123 characters.
(2) Multimedia icon that can be selected to view a ShowMe(TM) How video demonstration of the psupply.mpg file.	(2) Existing context assumed. Gives the name of the movie file rather than the movie title. Short alternative text: 103 characters.

EXAMPLE 13–10 Animations

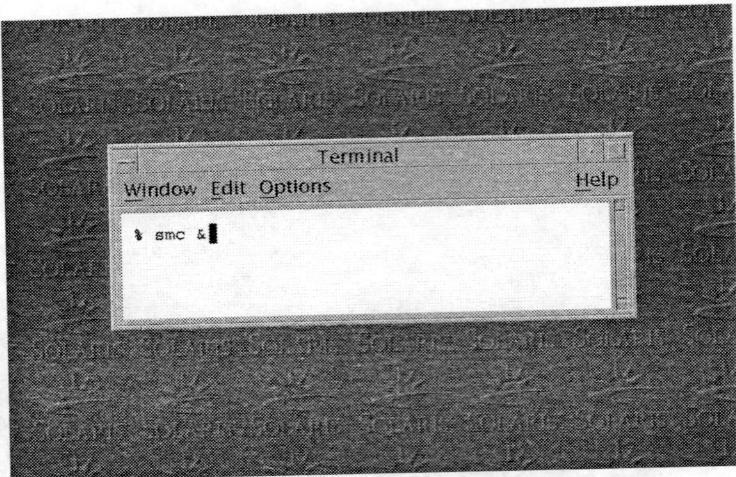

See the full animation and its alternative text at `http://webhome.corp/pubs_info/Int/alttextexamples.html`.

Alternative Text	Notes
(1) The animation introduces the Solaris Management Console features.	(1) Existing context assumed. Because these animations include an audio track that reads the onscreen captions, long alternative text is not needed. Short alternative text: 66 characters.
(2a) The animation introduces the Solaris Management Console features. (2b) The animation includes screen captures for the following steps. Step 1 shows the smc command that starts the Solaris Management Console. Step 2 shows the Navigation window, which displays the set of available tools. Step 3 shows the toolbox, which groups the tools into a consistent, user-friendly hierarchy. Step 4 shows the View pane, which provides icons that you click for more details. Step 5 shows details about a clicked icon in the View pane.	(2) No existing context assumed as no audio track that reads the onscreen captions is available. Could not describe the steps in the animation without exceeding the 150-character limit, so long alternative text was added. (a) Short alternative text: 66 characters. (b) Long alternative text: 462 characters.

EXAMPLE 13–11 Simulations

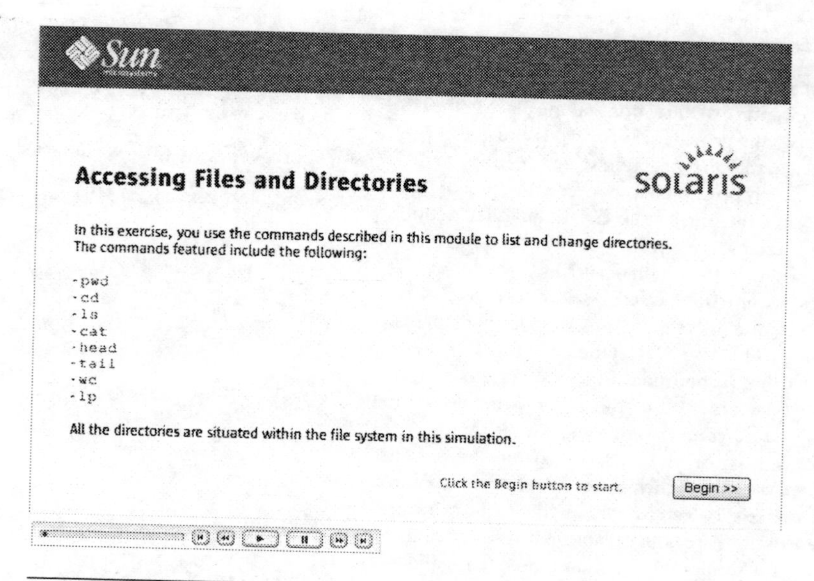

Alternative Text	Notes
(1) This simulation is of an onscreen lab exercise.	(1) Existing context assumed. Short alternative text: 51 characters.

EXAMPLE 13–11 Simulations *(Continued)*

Alternative Text	Notes
(2a) This simulation is of an onscreen lab exercise. (2b) Each frame of the simulation shows a dynamic screen capture of the actions taking place in the lab exercise, and provides explanations and questions related to the current action screen. You are periodically prompted to answer a question or perform an action. The explanations and directions for each frame are printed below, as well as a brief description of the action occurring on each frame. At times, you might be asked to select the correct answer to a question or to type text on the command line or in a login box. If you select the wrong answer, you will receive the error message: Incorrect. Please try again. If you type the wrong command or login information, you will receive an error message similar to the following: Incorrect. Check your syntax for use of proper case. If you click areas of the screen that are not used in this exercise, you might get the message: These options are not available in this simulation. Frame 1 describes the commands used to list and change directories, which are the pwd, cd, ls, cat, head, tail, wc, and lp commands. This frame also instructs you to click Begin to start the exercise. Frame 2 shows a terminal window with the dollar-sign prompt. Under the terminal window, you are directed to display your current working directory. Type the pwd command at the prompt.	(2) No existing context assumed. No audio that reads the captioning is included. Could not describe the steps in the simulation without exceeding the 150-character limit, so long alternative text was added. (a) Short alternative text: 51 characters. (b) Long alternative text: 1318 characters.

◆ ◆ ◆ CHAPTER 14

Documenting Graphical User Interfaces

A *graphical user interface* (GUI) provides the user with a visual way of interacting with a computer and its applications. The main purpose of a GUI is to make the activities that are involved in doing a task simple and quick. Common GUIs require a person to use a mouse or some other pointing and selecting device. GUIs are typically built around *windows* and *menus*, which provide a simple means of manipulating files and directories. However, the term "graphical user interface" also includes web browsers.

This chapter discusses the following topics:

Using GUI Terminology

Most computer users today are familiar with GUIs. However, you might still need to explain to novice computer users the basic concepts of the GUI and how to work in it.

Once you have established how to work in a GUI, write about the task the user wants to accomplish. You need not describe the detailed steps required to initiate the task. For example, a user rarely chooses menu options or clicks buttons with the goal of displaying other menus or dialog boxes. Rather, users want to accomplish a task that is activated by the menu option or button.

Writing About GUIs

When writing about GUIs, follow these guidelines:

- Provide only the essential details or information that the user needs to know to accomplish the task.
- Distinguish windows, dialog boxes, and menus with their proper names. Avoid repeated use of the same name when the meaning is clear.

  For example, say you have told the user to choose Print from the File menu and the Print window is displayed. You do not need to keep saying "in the Print window" if that is the window in which you want the user to work.
- Show field names and menu option names as they appear in the GUI.

  Consider adding initial capital letters even if the word is not capitalized in the interface to increase the readability of running text.

  Incorrect: Select Prompt for credentials for every server for maximum security.
  Correct: Select Prompt for Credentials for Every Server for maximum security.
- Using special text formatting such as an alternative font or bold type for GUI element names such as menu names, menu options, or field names can be confusing and distracting, especially in documents with a large number of GUI procedures.
- Do not include a colon or ellipsis points associated with a field name or menu option when mentioning the item in running text.
- Mention certain technical GUI distinctions only when necessary.

  For example, mention that a menu is a *pop-up menu* or a message is a *status message* only if that information is necessary. This usage might be appropriate, for example, if you have multiple types of menus with similar names.

Note – For legal reasons, be careful when providing examples in GUI documentation. See "Protecting Confidential Information in Examples" on page 200.

Writing About Mouse Actions

When writing about mouse actions, follow these guidelines:

- The plural for "mouse" is "mouse devices."
- The buttons on the mouse are referred to as *mouse buttons* to avoid confusion with control or command buttons in application windows.

  Right-click in an empty part of the workspace.
  Click the right mouse button and drag across the object.

- The indicator that shows where the mouse action occurs is called the *pointer* or *cursor*.

  Keep the user's attention focused on the screen by writing about the pointer, not the mouse.

  Incorrect: Drag the mouse so that the cursor is on a corner of the object.
  Correct: Place the cursor on a corner of the object.

Using Common GUI Terms

When writing about GUIs, reserve certain terms for specific activities. The following table lists common usages. The terminology of your GUI might require alternatives.

TABLE 14–1 GUI Terms

Verb	Action	Example
Browse	To examine an interface to locate information of interest. Do not use "explore."	The `prodreg` tool enables you to browse, unregister, and uninstall software in the PlirgSoft Product Registry. Browse the PlirgSoft Product Registry.
Check	Use "select" for the action of placing a checkmark or other indicator in a check box or radio button.	Select the Owner check box. (Some interfaces use "checkbox.") Select the Always option.
Choose	To open a menu or initiate a command.	Choose New from the File menu. Save your file, then choose Print.
Clear	Usage depends on your primary audience. Another common verb for the action of removing a mark from a check box or radio button is "deselect."	Clear the Set as Main Project check box. (Some interfaces use "checkbox.")
Click	To press and release a mouse button without moving the pointer. **Note:** If you are referring to a common button name (such as OK) or have established a button name used in the application, you can eliminate the surrounding phrase "the ... button."	Click the left mouse button. Click OK. Click the Update Actions button. Click the Glossary link. Click the Help tab in the wizard for more information.
Close	To close a window. Synonym for "dismiss."	Close the Add Users window.
Collapse	To click a turner icon (or similar GUI element) to hide subentries in the hierarchical structure you are viewing.	Collapse the view of the SCSI targets and disks under each disk controller.
Copy	To place a duplicate of the selection on the clipboard.	Copy the first figure in Appendix B.
Cut	To remove the selection from the current location and place it on the clipboard.	Cut the second entry in the list.

TABLE 14–1 GUI Terms *(Continued)*

Verb	Action	Example
Deselect	Usage depends on your primary audience. Use "deselect" for the action of removing a checkmark from a check box unless the software platform for which you are writing uses the term "clear."	Deselect the Generate Code check box. (Some interfaces use "checkbox.")
Dismiss	To close a window.	Dismiss the Add Users window.
Double-click	To click a mouse button twice quickly without moving the pointer.	Double-click the File Manager icon to open the program.
Drag	To move the pointer or an object by sliding the mouse with one or more buttons pressed.	Drag the pointer to draw a text box. Drag the icon to the upper left corner of the screen.
Expand	To click a turner icon (or similar GUI element) to view subentries in the hierarchical structure you are viewing.	In the Projects window, expand the Model node.
Highlight	Use "highlight" as a noun or adjective describing the appearance of selected text or an active button. Use "select" or "emphasize" as the verb form.	The selected text is highlighted. You can display the highlight in color.
Move	To move the pointer on the workspace by sliding the mouse with no buttons pressed.	Move the pointer outside the Mail window.
Navigate	To move through a web page. Do not use "explore" or "surf."	Use all three panes to navigate the directory, perform configurations, and create policies. Navigate through the browser interface as you would through a web page.
Open	To start or activate an application, or to access a document, file, or folder. To open an application window or browser window.	Open FrameMaker. Open the Samples document. Open the Add Accounts window. Open the Display Users page.
Paste	To place the clipboard contents at the insertion point.	Paste the figure into Chapter 3.
Point	To move the pointer to a specific location on the screen by moving the mouse with no buttons pressed. After the user is familiar with mouse techniques, you do not have to use this term. You can simply write "Click the Trash icon."	Point to the Trash icon and click to select it.
Press	To push down on and then release a keyboard key.	Press Return. Press Ctrl.

TABLE 14–1 GUI Terms (Continued)

Verb	Action	Example
Release	To let up on a keyboard key or mouse button to initiate an action.	Drag the file and release the mouse button.
Select	To highlight an entire window or data in a window, or to pick something from a list. To describe picking something from a menu, use "choose." Use "select" as the verb and "highlight" as the noun. Use "select" for the action of placing a checkmark in a check box.	Select the Compose window. Select the second sentence. A yellow highlight appears around the sentence. Select the file name in the list. Select the Owner check box. (Some interfaces use "checkbox.")
Size	To enlarge or reduce the size of a window.	Size the Text Editor window so that you can display more characters per line.
Start	To activate a browser.	Start the browser.
Type	To enter or specify information in a field.	Type a name and a description.

Writing About Windows, Dialog Boxes, and Menus

A graphical user interface includes the basic elements in which the application displays text and the user interacts with the application: windows, window controls, dialog boxes, and menus.

For examples of how to write procedures for GUI applications, see Chapter 7, "Writing Tasks, Procedures, and Steps."

Window Elements

A *window* is the primary, rectangular area in which application elements are displayed.

The elements in a window vary from application to application. The names of window elements appear in the document's default font. The following table describes some common window elements.

TABLE 14–2 Window Elements

Element	Description
Background	Bordered rectangle in which the application displays its data or the user enters data
Control area	Region where controls, such as buttons or settings, are displayed

TABLE 14–2 Window Elements (*Continued*)

Element	Description
Footer	Region in which the application displays status and error messages, or state and mode information
Header or title bar	Region in which the application displays a title as a long-term message
Icon	Pictorial representation of a base window
Pop-up window or secondary window	Window that provides related or secondary information

Window Controls

Controls in windows enable the user to perform actions. The names of controls appear in the document's default font. For controls such as sliders, radio buttons, and check boxes, name the control type only when you need to distinguish it specifically. Some common window controls are described in the following table.

TABLE 14–3 Common Window Controls

Control	Description	Example
Button	Small area within the main window on which the user clicks to execute commands (command button), display pop-up windows (window button), or display menus (menu button). **Note:** If you are referring to a common button name (such as OK) or have established a button name used in the application, you can eliminate the surrounding phrase "the ... button."	Click OK. Click the Replace Data button.
Check box (some interfaces use "checkbox")	A yes or no, or on or off, control. When a setting is selected, a visual indicator appears in the square check box.	Select the No Markers check box.
Gauge	A read-only control, usually at the bottom of a window, that shows the percentage of use or the portion of an action that has been completed.	The gauge in the status bar shows the progress of the document composition.
Radio button	A yes or no, or on or off, control. Usually, only one radio button in a group can be selected.	Select the Add User option.
Resize corner	Control that enables the user to size the window without changing the scale of the contents.	Drag the resize corner to change the size of the window.

TABLE 14–3 Common Window Controls (Continued)

Control	Description	Example
Scrollbar	Control that moves the view of the data that is displayed in the window.	Scroll to the top of your document.
Slider	Control that is used to set a value within a range and to give a visual indication of the setting.	Drag the slider to the desired saturation value.
Text field	Input area for text. If the field name includes a trailing colon, do not include the colon when referring to the field in running text. Text field names do not generally appear in an alternate font. Use monospace font for any specific text that you are instructing the user to type into a field.	Type your company name in the Organization field.

Dialog Boxes

A *dialog box* is a pop-up window in which the user provides information or issues commands. (In some GUI applications, a dialog box is called a *secondary window.*) Use the full term "dialog box" to refer to dialog boxes rather than "dialog."

Keep in mind that a user does not select menu options or buttons "to display the dialog box" but to do a task through the dialog box.

Incorrect:

1. **Choose Add Users from the Actions menu to display the Add Users dialog box.**

Correct:

1. **Choose Add Users from the Actions menu.**

   The Add Users dialog box appears.

Menus

A *menu* or a *submenu* is a list of application options.

If you are writing for novice users who might be unfamiliar with GUIs, use step-by-step instructions when you first write about choosing items or settings from a menu.

Menu option names appear in the document's default font and do not receive any special typographic styling. Also, even if a menu option includes ellipsis points in the GUI, do not include the ellipsis points when mentioning the option in text.

Here is an example of detailed steps that use menu options.

To paste text:

1. **Press mouse button 1 on the Edit button in the menu bar.**

   The Edit menu is displayed.

2. **Drag the pointer to the Paste item and release the mouse button.**

   The contents of the clipboard are placed at the insertion point.

After you have explained how menus work, streamline the process by saying "Choose Paste from the Edit menu" or "From the Edit menu, choose Paste."

Alternatively, you can use an arrow symbol to show the menu selection, for example, "Choose Edit → Paste." To insert the arrow symbol, use the correct symbol or character entity in your authoring environment. For example, if you are writing in HTML or the DocBook DTD, use the character entity `rarr`. However, "Choose Edit → Paste" is merely an example. Instructions in the GUI that you are documenting might use a different style for menu selection, which you would mirror in your document.

You can also use an arrow symbol to show a submenu selection. For example, "File → New → Templates."

Writing About the Web

The World Wide Web has attracted a diverse community of web designers and writers, with no predominant editorial style. Further, no solid consensus has been reached on industry-standard terminology for browsers or navigation. This section provides some guidelines for usage.

Web Terminology

Use the following terminology when writing about the web or a browser interface:

- Use "browser" rather than "web browser" or "web browser window" when referring generally to web browsers or to the main or first web browser window.
- Use "page" rather than "window" when referring to a web page.
- Use "user interface" when you are distinguishing between a "browser interface" and another interface. For example:

  The system provides two user interfaces for accessing the configuration and monitoring software:

  - A browser interface for running the graphical user interface on any host networked with the system. The browser interface is the primary interface for configuring, managing, and monitoring the system.

  - A command-line interface (CLI) for running commands interactively in a terminal window and for writing scripts to automate certain administrative tasks.
- Assume that URL is pronounced "you-are-ell" and use the appropriate related article ("a URL").
- Use a lowercase "w" except in the expression "World Wide Web" or in a product name.

  For more information, see the Plirg web site at `http://www.plirg.com`.
  The World Wide Web has fundamentally changed information retrieval.
  This section describes how to install the PlirgPak Web Console software.

Referencing URLs

Use monospace font for URLs.

Note – For information about legal requirements for linking to a third-party web site, see "Referencing External Web Sites" on page 194. For guidance about writing links, see Chapter 6, "Constructing Links." For information about referring to online articles or journals, see "Citing Electronic Sources" on page 81.

When introducing URLs, follow these guidelines:

- Make references to URLs as simple and direct as possible.

  Readers no longer need to be told to type a URL. For example, you do not need to write detailed instructions such as "Point your browser at the following URL."

  Incorrect:

  You can find further information by typing the following URL: `http://www.plirg.com`.

  Correct:

  Further information is available at `http://www.plirg.com`.

  Incorrect:

  Open the following URL in your browser: `http://www.plirg.com`.

  Correct:

  Go to `http://www.plirg.com`.

- Precede a URL with "at" or "to" when a preposition is needed.

  Incorrect:

  Check for updates on `http://www.plirg.com`.

  Correct:

  Check for updates at `http://www.plirg.com`.

  Correct:

  To check for updates, go to `http://www.plirg.com`.

- Provide explanatory text about the web site before giving the URL.

  A URL is acceptable at the end of a sentence. Readers no longer confuse end punctuation with spelled-out URLs.

  Incorrect:

  Go to `http://www.plirg.com` for resources for developers and service providers.

  Correct:

  For resources for developers and service providers, go to `http://www.plirg.com`.

  Correct:

  PlirgSoft offers resources for developers and service providers at `http://www.plirg.com`.

When placing URLs in text, use the following guidelines:

- If the URL is long, place the URL on its own line, introduce it with a colon, and do not put a period at the end.
- If the URL is short, weave it into the sentence.
- If your document provides two or more URLs in the same paragraph, place them in a bulleted list.

If you need to refer to an extremely long URL, use one of the following techniques:

- Provide the base URL and an indication of which links to click to get to the specific page.

  Awkward:

  For more legal information, see:
  `http://plirglegal.north/Legal_Dept/plirglegalweb/ legal_orgchart.html`.

  Clearer:

  Click Organization Chart on the main Legal web site at `http://plirglegal.north`.

- Substitute text for the URL in print and use the same text online as a link.

CHAPTER 15

Creating Screencasts

A *screencast* is a digital recording of computer screen output, also known as a *video screen capture,* often containing audio narration. This chapter explains the general process for developing and recording a screencast. It also provides a link to an example of a screencast. The chapter covers the following topics:

Screencast Overview

A screencast is most often used to demonstrate computing procedures and software functionality and includes audio narration (what the audience hears) that corresponds to the onscreen action (what the audience sees). A screencast has a high impact on the audience and can serve as a supplemental or replacement piece for more traditional customer-facing materials. Additionally, internal transfer of information (TOI) sessions can be captured in screencasts for on-demand playback.

Your audience is likely to view and listen to a screencast in linear fashion, that is, from start to finish. Thus, a screencast might well be suited for demonstrating product functionality or interface features, for example, but might not be as useful for long, complex procedures.

The duration of a screencast depends on the subject and topic. A screencast can range from 2–20 minutes, or longer. However, while duration is not limited, aim for a short, concise screencast that captivates the audience while meeting their needs. Provide information t qhat is needed for the particular topic and nothing more. In addition, rather than creating one 30-minute screencast that covers multiple topics, consider creating three 10-minute screencasts organized by topic, task, or both.

For an example of a screencast, go to `http://webcast-west.sun.com/interactive/08D12331/`.

Developing a Screencast

The following process is intended to help you develop screencasts. Most screencasts will benefit from the use of a storyboard, which contains narration (spoken content) and corresponding media segments (visual content). The target audience usually determines whether a storyboard is needed and the depth of the storyboard contents. A TOI that is available only internally might not require the amount of planning required by a storyboard. A screencast that will be hosted externally or presented to an external audience should be carefully planned and reviewed.

If you intend to develop multiple screencasts, consider creating a storyboard template. Split the template into numbered segments that contain the narration and the description of the corresponding visual content.

Note – The four reviews mentioned in the next two sections are explained in "Screencast Review Cycle" on page 286.

Developing the Storyboard Content

1. Meet with your media production contact, if applicable, to discuss your concept for a screencast.
2. Create a detailed outline of the content that the screencast will follow.
3. Using the storyboard template, fill in the storyboard with the content from your outline.
4. When appropriate, describe requirements for the visual content of the screencast by using the storyboard template.

   Note specifically what should be shown onscreen during the corresponding narration.
5. Discuss the media requirements with your media production contact to agree on the requirements.
6. Have stakeholders (media, technical, and editorial) review the draft storyboard for approval of proposed content and media. (*Review One*)

Creating and Deploying the Screencast

1. Using the storyboard, complete a dry run of the screencast to make sure that all content is accounted for and works as intended.
2. Edit the storyboard as necessary and have stakeholders review it for final approval. (*Review Two*)
3. Copy the narration from the storyboard into a separate document to create a *script*, and give a copy of the script to your media production contact for the recording.

   This script contains only narration and will be used by the narrator to record the narration.

4. Using the storyboard and script, record the screencast and narration.

   If voice talent is used instead, see your media production contact beforehand to coordinate the talent and to schedule a recording time.

5. Have stakeholders review the final screencast. (*Review Three*)

6. If changes are required, update the storyboard, edit the screencast as appropriate, and have stakeholders review and approve the updated screencast. (*Review Four*)

   Repeat this step until the screencast is approved for the final time.

7. Deploy the screencast or work with your media production contact, who can assist with this process based on your audience and needs.

Note – Because this process is iterative, build extra time into your schedule. For example, parts of the storyboard might have to be rewritten, parts of the narration re-recorded, or some media segments reworked. Some of these tasks might have to be completed more than one time.

Having stakeholders conduct reviews throughout the process helps to ensure a successful screencast.

Storyboard Overview

Using a storyboard can help streamline the process of both creating the content and ensuring that a screencast meets the desired objective. By using a storyboard, you will find gaps and flow issues in your screencast more easily and quickly. Finding and fixing problems when developing the storyboard is better than having to do so later in the screencast process.

Definition of Storyboard

A *storyboard* is divided into segments and contains information that connects the narration with what happens onscreen. A storyboard can help you outline exactly what you want to do, how to do it, what to say, and when to say it. In addition, a storyboard enables all involved parties to review the content and generate consensus prior to production of audio and visual segments. As a result, the likelihood of required changes post-production is reduced.

For an example of a storyboard template, go to `http://mediacast.sun.com/users/damon./media/screencast-storyboard`.

Developing a Storyboard

When developing the storyboard, keep in mind the following suggestions:

- Use the storyboard template.

  A storyboard is recommended if the screencast will be translated, if you must comply with accessibility requirements (for example, if the screencast is part of a training deliverable or might be used by the federal government), or both.
- Create narrative text that appropriately addresses each step in a procedure.
- Avoid unnecessary narration gaps, for example, where no narrative matches what is happening on the screen.
- Include textual description specific to areas that can benefit from highlights, callouts, or the display of graphical content (for example, diagrams).
- Use slides as appropriate.

  For example, you might want to list the benefits of a feature that you are about to demonstrate. Or, you might want to use slides to display graphical content, for example, a photograph or a diagram.
- Note areas that can be time-compressed, for example, long software installation steps.
- Submit the draft storyboard to media, technical, and editorial stakeholders for review.

  The stakeholders should also review the storyboard at various stages throughout the screencast production process. See "Screencast Review Cycle" on page 286.

Writing Narration for Screencasts

This section provides writing, terminology, and word choice guidelines to help you develop narration (spoken content) that is well written, appropriate for recording, and easily translated. This narration will ultimately become the script that the narrator will use.

Note – For additional writing guidelines, see Chapter 3, "Writing Style."

Narration Writing Guidelines

Follow these narration writing guidelines:

- Determine the nature of the narration.

  The objectives and target audience (for example, user compared with developer, internal compared with external to the company) usually determine the contents.

- Introduce what the screencast is about in the first 30 seconds.

  You want to motivate people and keep their attention. For example, an introduction might be written as follows: "Welcome to PlirgPak! This demonstration provides a quick introduction to the PlirgPak workflow by walking you through the creation of a simple 'Hello World' console application. After you have viewed this demonstration, you will know how to create, build, and run applications in PlirgPak."

- Be aware that language that works in slides or in print does not necessarily work as well when spoken.

  For example, the following text sounds awkward when spoken: "Bullet 4: To see what is new with the PlirgSoft Operating System (PlirgSoft OS), go to `http://www.plirgware.com/docs/div/1531.1`."

- Write in a conversational tone or one commensurate with the content.
- Write in a way that assists translators.

  Because the narration and onscreen text might be translated, use correct and consistent capitalization, punctuation, terminology, and font conventions.

- Write simply, directly, and accurately.

  Your audience can lose attention when listening to long sentences or to too much information.

- Avoid the use of future tense and passive voice, unless doing so is required by the context.
- Avoid sexist language and humor.

  Typically, the use of humor is inappropriate, especially for external and international audiences.

- Use Notes and Cautions as necessary.

  See "Notes, Cautions, and Tips" in Chapter 2, "Constructing Text."

- Limit the use of lists.

  Bulleted lists and numbered lists are awkward when spoken. Make the text more conversational.

  Incorrect:

  In the first bullet,... In the second bullet,...

  Correct:

  One consideration is...

  A second consideration is...

- Make the text describing steps conversational.

  Incorrect:

  Step 1..., Step 2...

  Correct:

  First, you... Next, you...

  In the first step,... In the second step,...

- Provide explanatory text and visual cues.

  Explain what is to happen after each step. For example, if a window opens as a result of a step, state that the window opens, and refer to the window by name.

- Describe GUI instructions clearly.

  For example, in print you might see "Choose File→New Project." You would write this step in narrative text as "Choose New Project from the File menu." See Chapter 14, "Documenting Graphical User Interfaces."

- When proceeding to the next procedure, cue the audience.

  Incorrect:

  Next, we need to define an interface for our BPEL process by using WSDL.

  Correct:

  The next procedure is to define an interface for the BPEL process by using WSDL. This procedure requires several steps.

Terminology Guidelines

When writing the narration, follow these terminology guidelines:

- Make sure that the name of the product or service is not used repeatedly.

  Instead, use "product" or "service" or "it."

- Do not use first person pronouns.

  The use of "I" and "we" comes across as awkward. Instead of saying "I will now...", say something like "The next step to perform is...." The use of the imperative mood in a "recipe" style is more professional.

  You can say "As you can see...", but keep the use of personal pronouns to a minimum. Some other examples include the following:

  Incorrect:

  If I resize the partition...

  Correct:

  Resizing the partition...

Incorrect:

In this presentation, we will show you how to...

Correct:

This presentation shows how to...

Incorrect:

Next, let's create...

Correct:

Next, you need to create...

- Minimize the use of "your."

  Only use "your" when not using it sounds awkward. For example, write "backing up your laptop," not "backing up the laptop."

- Avoid using indefinite terms.

  When proceeding to a new segment in the storyboard, repeat any nouns instead of using "it" or "this" by itself. For example, in the first and second segments, write the following:

  Incorrect:

  The standard module... You deploy it...

  Correct:

  The standard module... You deploy the standard module...

- Do not use "please" or "simply" or other unnecessary terms.

  Such terms take up "verbal space."

- Use "now" sparingly.

  Consider whether "next" would be more appropriate. For example, write "Next, invoke the...", not "Now, invoke the...."

- Do not use wording such as "on the previous screen."

  Instead, write "as previously discussed." Likewise, do not use "on the following screen." In the segment for the following screen, write, for example, "Next,..." or "On this screen,...."

- When choosing one of multiple valid options, use wording such as "for now, assume" to indicate the choice.

  Incorrect:

  We will choose option 4...

  Correct:

  For now, assume you are using..., so you would choose option 4.

- Use technical terms consistently.

  For example, use "terminal window" rather than "terminal" when you need to use this term. Also, use appropriate verbs and terms to describe the interface and how to interact with it. See Chapter 14, "Documenting Graphical User Interfaces."

- Limit the use of abbreviations and acronyms, especially in external works.

  Some abbreviations and acronyms sound awkward when spoken. For example, say "operating system" (not "OS") and "high availability" (not "HA").

  Common, nonjarring abbreviations and acronyms are acceptable if appropriate for your audience. For example, CD, NFS, NIS, LDAP, I/O, PROM, GUI, DIMM, ID, USB, and LED. However, write these terms phonetically so that the narrator will pronounce them correctly.

  Also, do not use abbreviations for measurements. For example, use the complete term "megabytes" instead of the abbreviation "MB."

- Write troublesome terms phonetically.

  For example, "." is usually pronounced "dot," not "period." A number such as 256 is pronounced "two-hundred fifty-six."

- Try to provide the person who is recording the narration with a separate document that has pronunciation guidelines for terms, abbreviations, and acronyms.

- Do not spell out command syntax.

  Instead, summarize the command, perhaps saying the command name and the options that are being used with it. The audience will see what is typed onscreen, so you do not have to say, for example, `chmod -777 /myfs`. Just write about using the `chmod` command to change the file permissions.

- Do not spell out URLs.

  Instead, make a general reference unless the URL is short and easy to pronounce, such as `www.plirg.com`. For example, write "For more information, go to the URL onscreen." Then, highlight the URL onscreen by placing a box around it. Often, a URL appears at the end of a screencast so that people can refer to it later.

  If you are referring to more than one URL, say, for example, "Go to the first URL for...", "Go to the second URL for...."

- Make cross-references clear and refer to URLs when appropriate.

  Incorrect:

  For more information, see the Technical FAQ on the download site.

  Correct:

  For more information, see the Technical FAQ on the download site, available at the URL shown onscreen.

- If you need to use a date, be specific.

  For example, instead of saying "last fall," say "in September 2009." If the screencast will be available for more than a year, listeners could be confused by a vague reference.

See also Appendix B, "General Term Usage," for a list and brief explanations of how to correctly use terms that writers and editors frequently look up when working on documentation.

Word Choice Guidelines

Follow these guidelines to avoid wording that makes translators and your audience uncertain of your intended meaning:

- Use American English spelling and punctuation guidelines.

  One standard dictionary to use as a resource is the *Merriam-Webster Collegiate Dictionary*, which is available online at `http://www.m-w.com`.
- Do not use foreign terms.

  Even though some foreign terms are listed in U.S. English standard dictionaries, do not use them. Nonnative English speakers might not understand foreign terms such as "vis-à-vis," "via," and "vice versa," or Latin abbreviations such as "i.e.," "e.g.," and "etc." In addition, when the screencast is translated, no equivalent terms exist for foreign terms in target languages.
- Avoid using contractions that are difficult to translate.
- Use terms consistently.
  - Avoid using terms that can have several different meanings.

    For example, the word "system" can refer to an operating system (OS), a combination of OS and hardware, or a networking configuration. If you do use such a term, make sure that you define it and use it consistently.

    Likewise, synonymous terms can be troublesome for a translator. The words "show," "display," and "appear" might seem similar enough to use interchangeably. However, a translator might think you used the different words deliberately for different meanings and might interpret the text incorrectly.
  - Use uppercase and lowercase letters consistently in similar elements.
  - For command names or other computer terms, include a phrase such as "the ... command" or "the ... function" on first reference. Including the surrounding article and term can clarify references.
- Avoid informal language and styles such as slang, irony, metaphors, idioms, and nonstandard colloquialisms.
- Avoid jargon.

  If a term is not listed in a standard dictionary or a technical source book, do not use it.

Recording Narration

This section contains guidelines for recording narration that is informative, appealing to the audience, and of high quality. You might or might not be the narrator for your screencast. Voice talent might be used.

Reviewing the Storyboard Before the Narration Is Recorded

Before recording the narration, follow these guidelines:

- Read the narration aloud while actually interacting with the software.

  This read-through is perhaps *the most valuable task* that you can perform when creating the screencast.
- If something sounds awkward, rewrite that portion of the narration and update the storyboard accordingly.

  Much of the review process is about evaluating the narration to determine whether it sounds right. Also, make sure that you have not omitted any text.
- Test procedures to make sure that they work and that they include all steps.
- If long pauses occur while waiting for something to happen onscreen, determine whether you need additional narrative text.
- Check the synchronization between the narration and what is happening onscreen.

  Make sure that the narration does not get too far ahead of or behind the visual content.
- Practice reading the narration until you are comfortable with it.

  Correctly pronounce acronyms, product names, and technical terms. Incorrect pronunciation of technical terms is a common reason for re-recording.
- Have technical and editorial stakeholders approve the narration before it is recorded.

Note – Using a word processor application, copy all narration (spoken content) from the final, approved storyboard into a new document. This will become the script that the narrator will use to record the narration. For readability, do the following:

- Double-space the lines of text.
- Use a large point size.
- Do not hyphenate words at the end of lines in the script.
- Do not break noun phrases, prepositional phrases, and verbal phrases at the end of lines in the script.

  For example, do not break "the installation procedure," "at the beginning," or "can be installed" at the end of lines.

Recording the Narration

Follow these guidelines when recording the narration from the script:

- Use a conversational voice.
- Avoid speaking in a voice that is too loud or too soft.

  Use a solid and consistent tone of voice, but emphasize major points by changing your tone and speaking style.
- Enunciate, and speak clearly and slowly.
- Try to avoid distracting speech habits.

  For example, try not to say "um" and "okay." Also, avoid awkward pauses.
- Sound interested, not bored.

  You do not want to sound like you are reading the script.
- Do not have too many people with various accents speaking in a single screencast.

  However, clarity in delivery is more important than the accent of the narrator.
- Try to imagine that an audience is present.

  You are unlikely to have an audience while you are recording the narration.
- Do not flip pages of the script while you are speaking.

  Lay out single-sided pages of the script in front of you. Then, take breaks from recording when you need to move the pages or flip them, or when you need to pause.
- Do not spell out command syntax.

  Instead, summarize the command, perhaps saying the command name and the options that you are using with it. The user will see what you type onscreen, so you do not need to provide details about the syntax.

Screencast Review Cycle

Throughout the screencast development process, all stakeholders, including media, technical, and editorial, should participate in several required reviews. Multiple reviews help to ensure that the project is on track and that the next stage of development is successful. If reviews are skipped, the final product might contain inaccurate information and require changes, thus delaying the release. The following table explains what to look for at particular stages of screencast development.

TABLE 15–1 Screencast Review Stages

Review Stage	When	Review Elements
Review One	After initial draft storyboard is complete	■ Correct content and product names ■ Relevancy of material ■ Completeness of steps
Review Two	After final storyboard is complete	■ Correct content and product names ■ Relevancy of material ■ Completeness of steps ■ Tool has not changed since storyboard was created ■ Consistent capitalization, punctuation, terminology, and font conventions
Review Three	After screencast has been produced	■ Completeness of steps (per storyboard) ■ Synchronization between the narration and visual content ■ Correct usage and pronunciation of terms ■ Tone ■ Pacing ■ Tool has not changed since screencast was produced
Review Four If changes are requested as a result of Review Three, a follow-up review should take place.	Each time that changes are made to screencast	■ Required changes are implemented ■ Storyboard and script have been updated as appropriate Make sure that the final storyboard and script match the final screencast. ■ Synchronization between the narration and visual content ■ Correct usage and pronunciation of terms (if narration has been adjusted) ■ Tone (if narration has been adjusted) ■ Pacing (if narration has been adjusted) ■ Tool has not changed since screencast was produced

◆ ◆ ◆ CHAPTER 16

Using Wikis for Documentation

A *wiki* is a web site that provides an easy way to collaboratively create and edit interlinked web pages by using a simplified markup language. Wikis are a way to publish online information and to obtain community input. This chapter provides some guidelines for technical communicators who are using or plan to use wikis to present their content.

Note – This chapter does not cover areas such as what types of information are best presented in wiki format, advantages or disadvantages of the wiki format, or how to determine whether your information should be delivered in wiki format. The decision of whether to use a wiki to present your information should be determined by your audience, your product, and other considerations.

This chapter covers the following topics:

Wikis and Collaboration

The key advantage of the wiki model is the capability for audience interaction. Your audience can comment on and contribute to the wiki content. Wiki usage for technical documentation ranges from limited internal posting of content for technical review to the full presentation of external product documentation with the capability for community contribution and comments. (A "community" is defined as group of people involved in or interested in a documentation project, potentially including technical writers, editors, engineers and support staff, marketing and sales staff, customers, and users.)

Some ways to encourage wiki collaboration include:

- Establishing guidelines that clarify what is expected of community members and how responsibilities will be distributed.
- Developing guidelines that give wiki moderators a clear mandate about how to address problems as they arise. For example, guidelines can address how to handle edits that contradict each other.
- Providing documentation describing the wiki tools and processes.
- Spreading the message that the wiki is available and that contributions and comments are welcome. If potential community members include customers and users, consider communication media such as blog postings, social media sites, and the like.
- Asking community members to assist with well-defined tasks:
  - Assign one or more community members to periodically check for content holes and dead links.
  - Ask members with technical expertise to periodically review the content for technical accuracy.
  - Focus the community's attention for a limited time on specific content. Limit the content that needs attention to one page or to a group of related pages.

Value of Publications Expertise for Wikis

Even though wikis are often viewed as an informal way to present information, the expertise of technical communicators is still valuable to the development and maintenance of information presented in wiki format. The skills of technical communicators in creating quality content apply to wiki content as well as more traditional forms of delivery. Other ways in which the skills of technical communicators can be used are as follows:

- When initially developing a wiki site, technical communicators can help evaluate continuity, structure, and navigation from the perspective of a visitor to the site.
- Customer-facing content can be reviewed for house style, tone, and legal exposure.
- When contributors make content or organization changes, technical communicators can review the changes for coherence, completeness, and navigational issues, and make sure that information is easy to find.

Wiki Organization and Navigation Guidelines

Consider the following guidelines when organizing your wiki content:

- Limit page length.

  Books can be delivered on wikis, but they should be divided into smaller topics to keep pages relatively short. For example, Wikipedia sets a limit of 10 printed pages of main body prose for articles.

- Use consistent conventions for page names and URLs.

  A consistent page-naming convention and meaningful page names enable readers to more easily find information. Consistent page naming also enables you to create links more easily.

  Tip – Some wiki tools can also use consistent page naming conventions for certain types of automation.

- Analyze navigation.

  Structure the wiki carefully by including tables of content, topic lists, jump lists, task maps, or other advanced organizers or by using directional icons. Consider at least three levels of navigation: general wiki table of contents, section-level topic lists, and page-level topic lists.

- Decide whether to include navigation links on supporting pages to the following types of information:
  - Welcome or home page
  - Table of contents
  - FAQs
  - Help
  - Getting started or "read me first" information (including editorial guidelines)
  - How to contribute
  - Recently updated pages
  - Topic list or task list
  - New features
  - Documentation
  - Known problems and workarounds
  - Index
  - Contact information

Writing for Wikis

In most cases, you should follow the same guidelines when writing content for wiki presentation as you do for any other presentation medium. This section provides some additional writing guidelines and tips, and some terminology guidelines.

Wiki Writing Guidelines

Follow these guidelines when writing for wikis:

- Follow your company's editorial style. If your wiki content is open to outside contributors, consider providing a link to your company style guide or to a briefer external editorial style sheet to ensure consistency.
- If the wiki provides the official documentation for your product, maintain the same professional tone as for product documentation in any other medium. Consider using blogs for less formal information presentation or interaction.
- If your wiki is available to international audiences, consider the issues and guidelines presented in Chapter 8, "Writing for an International Audience."
- Follow the relevant guidelines in Chapter 5, "Online Writing Style," and Chapter 6, "Constructing Links."

Wiki Terminology Guidelines

Wikis have attracted a diverse community of software designers and contributors, with no predominant editorial style. Further, no solid consensus has been reached on industry-standard terminology for presentation or navigation. This section provides some guidelines for term usage.

Avoid referring to wiki content modules in text. Just start writing about the section's topic. For example, write "To upgrade the software..." instead of saying "This section describes..." When you have to refer to a type of module, use the wording described in the following table.

Note – Use book-related terms to refer to information on the wiki when this is the clearest way to describe a collection of information. For example, use "document," "guide," "chapter," and "part." Avoid using the term "book."

TABLE 16–1 Wiki Terminology

Term	Definition	Examples
Site	Refers to the overall wiki	This site contains the official information about the PlirgWare 4.8 software.
Page	The current page or another page within the site	This page explains what's new in the PlirgSoft Security Center. The following pages provide an overview of the PlirgWare security features.
Section	A part of the current page	This section describes how to upgrade the PlirgWare software.
Topic	A general catch-all term when referring to a combination of pages or sections	After reading this topic, you should have a good understanding of the major features of the PlirgWare software and how to implement them.

Avoid the following terms when referring to a wiki page:

- Article
- Subpage
- Subsection

Wiki Visual Design Guidelines

Make your wiki visually appealing to help visitors find what they want and to encourage them to participate in its development. An effective use of visual elements such as colors, fonts, and graphics can give cues about the wiki's organization and navigation. The use of product logos and other graphics can personalize your wiki and make it compelling and unique.

Note the following general visual design guidelines:

- Consider using the following heading levels to make the distinction between levels clear: h1, h3, and h5. Sequential levels of headings can be difficult to discern in wikis.
- Use visual elements such as diagrams, icons, photographs, and other graphics, where appropriate.

Note – Be sure to provide alternative text for the graphics used in your wiki for the benefit of people using screen readers or text-only browsers. For more information, see Chapter 13, "Writing Alternative Text for Nontext Elements."

- Use icons as a navigation aid to help create visual interest and to improve usability by breaking up long columns or rows of text.

  Do not overuse icons or repurpose them beyond their intended scope. Keep an icon's definition consistent among wiki authors. When using icons in your wiki, accompany them with names that explain their purpose, as well as alternative text.
- Consider adding your product's logo at the same place on every page for additional context.
- Use color to add interest but avoid overuse that can overwhelm the text or the page.

  For example, you could use color for table backgrounds to differentiate rows and columns of information. You could also use color to group related pages on specific topics by assigning a color to each topic by using a small colored text box next to the topic's title and throughout the topic in key locations.
- Use templates to promote consistency in your wiki pages.

  A template serves as a generic placeholder for content that is inserted into a new page. A template can contain headings, sample text, boilerplate text, a navigational scheme (though not usually), and visual design elements. The notion of templates also applies to parts of pages, not only entire pages. For example, a template for a box that provides links to useful information could be helpful if this type of information will be included on multiple pages in your wiki.

  Templates are also user friendly in that they help contributors create pages more easily and more quickly. Certain types of shared information are most likely to draw contributions, so consider developing templates for the following material:

  - FAQs
  - Troubleshooting tips
  - Procedures
  - Examples
  - Customer best practices
  - Style sheets

CHAPTER 17

Glossary Guidelines

This chapter explains how to create a glossary for a technical manual. Most technical manuals introduce a number of new terms. A glossary improves the usability of a technical manual by simplifying a reader's search for definitions for these new terms. Before you create a glossary, consider its use and organization.

This chapter discusses the following topics:

Glossary Content

The terms that you select for a glossary must be important to the subject, with simple and concise definitions that are appropriate for the context. A glossary can contain some background definitions that are not defined within your book if these definitions enhance understanding of the subject. An example is units of measure.

A glossary can also include terms that have another meaning apart from technology. Examples are "front porch" and "portal." You might also include terms that are unique to the book.

Some terms are not considered appropriate for a glossary. These terms include the following:

- Commands from a programming language or from an operating system
- Window menu options
- Terms that a reader can find in a standard dictionary or an industry glossary

Finding Definitions

For terms that are specific to your particular product, try to obtain definitions from your technical subject matter expert, usually an engineer. For standard industry terms, first check a standard dictionary. If the term is not in the dictionary, check the reference books in Appendix E, "Recommended Reading."

Online glossaries are usually easy to access. You might find the definition that you need without much difficulty.

Caution – If you need to borrow definitions from another source, ensure that you rewrite the definitions so that you are not guilty of plagiarism. To inspire originality, define a term as if you were creating a definition for someone who is unfamiliar with the term or technology.

Ensure that your glossary is verified in a technical review, along with your main text.

Creating New Terms

If you need to create an entirely new term for a product feature or component (or provide feedback for such a term), follow these guidelines:

- Ensure that the new term cannot be read as obscene and that its meaning cannot be mistaken.
- Ensure that a translator can adequately translate the new term.

  If you are not sure about a term and its definition, consult localization.

Formatting a Glossary

Glossary page formats differ from book to book. You can see these differences especially in Internet glossaries, which use a variety of publishing software for their online presence. All glossaries do have some common characteristics:

- All terms are arranged alphabetically.
- Each term begins on a separate line.
- Each term is followed by its definition or expansion.

Some glossaries capitalize the first letter of each important word in an entry. Other glossaries capitalize only proper nouns in terms. Examples in this chapter follow the correct capitalization of the term as used in the book. Terms often follow the capitalization that is recognized by a standards organization, such as ANSI, IEEE, or ISO. The typeface, margins, and indents are determined by the templates and the publishing software that you use to create the glossary.

Terms for an International Audience

If your document or documentation set has a glossary, ensure that the language of definitions in the text is consistent with that glossary. If you are working with a translation or localization group, provide a glossary and a style sheet for your document.

When you are writing glossary definitions, remember the concerns of translators and nonnative speakers. This audience sometimes struggles to understand a new term in English in the process of rewording the term in another language. Therefore, be as complete as possible when you write a glossary entry. Try to include the following items:

- Grammatical use in a sentence, when a distinction is important

  For example, is the term a plural noun that is more commonly used in English as a singular noun?
- Source, location, and usage of the term

  For example, where is the term used: in software, a graphical user interface, help files, hardware, marketing material, web navigation, or legal documents?
- Sample sentence that shows the term's usage (optional)

In many instances, the term is so widely used that the particulars do not need to be carefully defined. But the value of the glossary increases as you add details for the international audience. For more information about internationalization, see Chapter 8, "Writing for an International Audience."

When to Include a Glossary

If you are documenting new technology, a glossary is almost mandatory. An audience that is unfamiliar with the topic can probably benefit from a major set of definitions.

To decide whether you need a glossary for your book, consider your content. If your content describes a new product with many new vocabulary words that you define within the text, include a glossary.

Alternatively, if the scope of your content does not extend to new material, you might not need a glossary. If you do not need to define many terms as you write, your audience might be familiar with the terminology.

If your book is a legacy book that has never had a glossary, add a glossary if you have time. The inclusion of a glossary especially applies to books that are undergoing major revisions. If you define many new terms within the text, you need to create a book glossary.

Writing Good Glossary Entries

Ideally, you define each term at least twice:

- At its first usage in text

  Give a brief but concise definition of the term.

- In the glossary

  Users expect a complete definition in the glossary.

The decision about whether to define a term in a subsequent chapter is at your discretion. You might define a term more frequently depending on context. Consider the following:

- If the page range is wide between the term's definition in one chapter and the subsequent use of that term in another chapter, you might provide a short, summary definition in the subsequent chapter.
- If the term is used in distinct contexts throughout the book, as in an overview chapter and in a task chapter, you might include the term's definition in both places.

The best time to write a glossary entry is when you first introduce an unfamiliar term in your text.

Introducing Glossary Entries in Text

When you write text, you frequently need to introduce terms to your audience. Define new terms and technical terms that are not listed in a standard dictionary the first time that those terms occur in text. Italicize terms when you first define them. Also, include these terms in a glossary. For example:

> Several configuration files are included with your package. A *configuration* file is a text-readable file that is used to set up or configure a part of the system.

The use of italic indicates that an entry for the word "configuration" appears in the glossary.

Tip – Some authoring environments for online presentation enable you to link a definition directly to the italicized term.

---

configuration (1) (n.) The way that you have set up your computer.

(2) (n.) The combination of hardware components that compose a computer system: CPU, monitor, keyboard, and peripheral devices.

(3) (n.) The software settings that enable various hardware components of a computer system to communicate.

Creating a Glossary Entry

Glossary style usually presents the glossary term in a simple, singular noun form. The part of speech that you use for the term demands that you use the same part of speech as a keyword in the definition. An example follows.

concatenate (v.) To string together two or more sequences, such as files, into one longer sequence.

node (n.) An addressable point on a network. Each node in a network has a different name. A node can connect a computing system, a terminal, or various other peripheral devices to the network.

Defining the Term

The definition, which immediately follows the term, does not restate the term. The restatement might be assumed, along with the unwritten words "is a" or "is," if the article is included. An example follows.

debugger (n.) A program that locates operational errors in another program. The debugger usually enables the developer to examine the malfunctioning portion of the program for bad data and to check operational conditions. See also debug.

decryption (n.) The process of converting encrypted data to plain text. See also encryption.

Ensure that the definition agrees in number with the term. The definition does not need to be a complete sentence.

Defining Multiword Terms

Defining a term that consists of multiple words, usually one or more modifiers and a noun, is similar to defining a single word. An example follows.

shell procedure (n.) An executable file that is not a compiled program. A shell procedure calls a shell to read and execute commands that are contained in a file. This procedure enables you to store a sequence of commands in a file for reuse. Also called a shell program or command file.

Defining Parts of Speech

Of all glossary users, translators and nonnative speakers can benefit the most from knowing the part of speech for a term. However, all users can improve their understanding of a term by knowing how to use that term in a sentence. The following table shows the abbreviations for parts of speech.

TABLE 17–1 Abbreviations for Parts of Speech

Abbreviation	Part of Speech
adj.	Adjective
adv.	Adverb
n.	Noun
v.	Verb

The following example shows two parts of speech:

format (1) (n.) The structure of data that is to be processed, recorded, or displayed.

(2) (v.) To put data into a structure or to divide a disk into sectors for receiving data.

Creating Multiple Definitions

When a term requires more than one definition, use a numeral that is surrounded by a set of parentheses and no other punctuation. Append a space to this text. Start a new paragraph for each definition, as the following example shows.

luminance (1) (n.) The generic flux from a light-emitting or light-reflecting surface. The subjective response to luminance is brightness.

(2) (n.) The specific ratio of color primaries that provides a match for the white point in a specified color space.

(3) (n.) The portion of a composite signal that carries brightness information.

Defining Acronyms and Abbreviations

Acronyms and abbreviations are alphabetized letter-by-letter, as described in "Alphabetizing a Glossary" on page 300. Occasionally, you need only give an expansion of the term. At other times, you need to provide a definition as well.

Because most users are familiar with just the acronym or abbreviation, spell out the meaning and give the definition immediately after the acronym or abbreviation. You can double-post the abbreviation and the expansion.

The following examples show entries with abbreviations and expansions:

SIP (single inline package) (n.) The packaging of an electronic component with all leads protruding from one side only.

SMPTE (n.) Society of Motion Picture and Television Engineers.

SNMP (Simple Network Management Protocol) (n.) The preferred network management protocol for TCP/IP-based internets.

When you use multiple definitions for an acronym or abbreviation, use a numbering system, as the following example shows.

BSD (1) (Berkeley Software Distribution) (n.) UNIX versions that were developed at the University of California, Berkeley. These versions have names such as BSD 2.7 and BSD 4.2.

(2) (block schematic diagram) (n.) A circuit board flowchart.

Note – If an acronym or abbreviation that is expanded in your book chapter contains initial capital letters, ensure that you repeat the capitalization in the glossary.

Using "See" and "See Also" References

When you refer to another term, use the word "See" to direct the reader to use another word. Do not define the term when you use "See."

solid model (n.) See surface model.

Use "See also" after you have defined a term to direct the reader to other terms with similar meanings. If you are referring to an acronym or abbreviation, use the short form. The following examples show two ways to use "See also."

composite drive (n.) A single logical drive that is composed of more than one physical drive. See also disk array, RAID.

`crontab` **file** (n.) A file that lists commands which are to be executed at specified times on specified dates. See also `cron`.

Alphabetizing a Glossary

The writer is responsible for arranging a glossary alphabetically. Most publishing software does not automatically alphabetize a glossary. Most glossaries are alphabetized by the letter-by-letter method. This method does not observe spaces between words. An example follows.

typefaces
type form
typescript
type size

Creating Online Links

For online use, your glossary definition can contain multiple online links. These links reduce the number of "See also" citations. In the example that follows, a set of underscores indicates online links. These links are not displayed as sets of underscores in online text.

namespace table (n.) The place where all _namespace_ information is stored, for use by the _classing engine_ as well as a _namespace manager_. Each namespace table consists of entries (rows) and each entry consists of a set of named attributes.

For more information about online writing style, see Chapter 5, "Online Writing Style."

Indexing Glossary Terms

Do not index the glossary. Because the glossary is arranged alphabetically or searched by keyword, a reader can find any definition quickly, with no need for index entries.

However, a reader might look for a glossary term in the index to determine how the term is used in the main text of the document. Consequently, some terms that are in the glossary might also appear in the index. These index entries should cite only pages that are in the main text.

CHAPTER 18

Indexing

An index is often a reader's primary information retrieval device. When readers search for a particular topic and find it referenced in an index, they are assured that the topic is covered in the document and their search has ended. When readers do not find a topic in an index, they might decide that the topic is not covered in that document and they might look elsewhere.

This chapter discusses the following topics:

What Is an Index?

According to *The Chicago Manual of Style*:

> A good index records every pertinent statement made within the body of the text. The subject matter and purpose of the book determine which statements are pertinent and which peripheral. An index should be considerably more than an expanded, alphabetical table of contents. It should also be something other than a concordance of words and phrases.[1]

The key word here is *pertinent*.

[1] *The Chicago Manual of Style*, 14th ed. Chicago: University of Chicago Press, 1993, p. 703.

Style and Format for Indexes

The following figure shows two examples of the indented indexing style. One example shows the index format for unnumbered chapters, the other example is for numbered chapters. The only differences between the two examples are the format of the page numbers and the use of "to" as a separator for page ranges in the book with numbered chapters.

EXAMPLE 18–1 Index Formats for Numbered and Unnumbered Chapters

Unnumbered chapter format example:

A

application architecture, 12
application gateway, **211**, 345–351
automounter facility
    See also mounting
    overview, 49
    setup, 51
        remote mounting, 52
        specifying subdirectories, 51

B

backing up
    file systems, 58
        dump command, 89–92
        dump strategies, 81

Numbered chapter format example:

A

application architecture, 1-2
application gateway, **4-18**, 5-76 to 5-81
automounter facility
    See also mounting
    overview, 2-12
    setup, 2-15
        remote mounting, 2-16
        specifying subdirectories, 2-14

B

backing up
    file systems, 3-1
        dump command, 3-34 to 3-37
        dump strategies, 3-26

Nested Index Entries

Most indexing styles use up to three levels of nested entries: *primary* entry, *secondary* entry, and *tertiary* entry. Each entry level is indented from the previous level. These three levels appear as follows:

primary entry
    secondary entry
        tertiary entry

The primary entry is the principal subdivision of an index. A simple primary entry includes the entry and a page number. A primary entry covering two or more page numbers is usually divided into secondary entries. Each secondary entry must bear a logical relationship to the primary entry. A secondary entry covering several page numbers can be further divided into tertiary entries. Each tertiary entry must bear a logical relationship to the secondary entry.

In an indented style, each secondary entry and each tertiary entry begins a new indented line unless there is only one secondary entry.

Least-Recently-Used (LRU) Ring functional
  description, 3-17
  elements of, 3-24

If an entry runs over the width of the column, it is indented in flush-and-hang style, that is, the first line is set flush and the rest of the entry is indented below it.

Page Number Style for Indexes

A comma and a space are inserted between the entry and the first page number. Subsequent page numbers are separated with a comma and a space. Page numbers appear in ascending order.

Do not include a page reference in the primary entry if it has two or more secondary entries.

functional description
  input block, 2-3 to 2-32
  introduction, 2-1
  output block, 2-33

Major Page References

In certain cases, you might want to identify a particular page as the main source of information for a given topic, especially if the topic cites two or more pages. You can identify the main page by marking the page number in bold.

application gateway, 211, **345–351**
⋮
dragging operations, 13, **37**, 114

Page Ranges

In a page range spanning several pages of a numbered chapter, separate first and last page numbers with a space, the word "to," and another space. For example:

screen adjustments, 1-6 to 1-12

In a consecutively numbered page range, use an en dash to separate page numbers. For example:

screen adjustments, 6–12

Special Typography for Indexes

The index is subject to many of the same typographic style conventions found in the text of the document itself. For example, if file names and commands appear in monospace font in your book, then they appear that way in the index.

`core` file, 16
`rlogin` command, 166

"See" and "See Also" References

"See" and "See also" cross-references require special consideration. For more information, see "How to Use "See" and "See Also" References" on page 316.

Capitalization for Indexes

Do not capitalize any word in an index entry unless the word is a proper noun, an acronym, or an abbreviation that is supposed to be capitalized. Use standard rules for capitalization.

subwindow button, 37
Sun Fire system, 34

Punctuation for Indexes

If an entry is followed immediately by page references, insert a comma between the entry and the first page reference, and between subsequent page references.

scrolling, 12, 16, 27

If you invert an adjective and a noun or noun phrase in an entry, separate them with a comma.

controls, window, 21

Use no punctuation between a primary entry without page numbers and subsequent secondary entries.

selection
    adjusting, 6
    extending, 59

Creating an Index

Indexing is an iterative process. Your first pass at an index is merely the foundation on which to build your final index. The first pass will be full of similar primary entries that you need to break into secondary entries. When you first develop your index, you might introduce spelling errors and have incomplete page ranges in individual entries. Such errors can be corrected during the editing phase. See "Refining and Checking an Index" on page 321 for possible solutions.

When an Index Is Needed

A document needs an index if it has 20 or more pages. This rule applies to any type of document, from a user's guide to a technical reference manual.

Time Required to Create an Index

Generally allow one full day of indexing for every 25 pages of text. A 100-page document might take four full days to index. An experienced indexer might require less time; a first-time indexer might require more time. Certain types of documents are much more difficult to index than others. Those documents require even more time than given in this guideline.

Deciding Which Parts of a Document to Index

The first decision you need to make when starting to create an index is to determine which of the following parts of a document to index:

- **Front matter.** Do not index the title page, copyright page, table of contents, and lists of figures, tables, and examples.
- **Preface.** Index the preface if it contains information about the subjects within the document and not just why the document was written. Topics you might index in a preface include prerequisite knowledge, other applicable documents, or document conventions.
- **Chapters.** Use the main body of the document as the source for most of the index entries.
- **Tables and figures.** Index topics within tables and figures if they are of particular importance to the discussion. Do not index items within tables that merely reproduce information already contained in the text. Including index references to the subjects of the tables and figures themselves is often helpful to a reader.
- **Footnotes.** Index footnotes if they contain information that expands on the information in the text. Do not index footnotes that merely document statements in the text.
- **Appendixes.** Index appendixes if they contain pertinent material omitted from the main body of the document. Do not index appendixes if they merely reproduce information already contained in the main body. A quick reference in an appendix, for example, is usually not indexed. Nor is a questionnaire. Worksheets in an appendix are indexed unless they merely repeat the main text.

- **Back matter.** Do not index terms in a glossary as long as they are explained elsewhere in the document. Do not index a bibliography or a reader comment form.

Selecting Topics to Index

When you take on the task of creating an index, you must first decide what the pertinent statements or topics are. A topic can be a single word, a phrase, or even a concept. A topic has no minimum or maximum size.

As you analyze a topic for inclusion in your index, decide whether the topic contains information a reader might expect to find in the index. If it does, create one or more index entries.

To determine whether a topic needs an index entry, analyze the topic for the following attributes:

- **Describes how to perform a task.** Tasks are the key subjects in certain types of documents, such as installation manuals.
- **Contains a definition of a term, acronym, or abbreviation.** Definitions are frequently the key to a reader's understanding of the document. An effective index makes it easy for a reader to find the definition or, similarly, an acronym or abbreviation. However, do not identify in an index definitions, acronyms, and abbreviations by glossary page numbers, but rather by text page numbers. See "Create Index Entries for Acronyms and Abbreviations" on page 311.
- **States a restriction, such as a Caution.** Awareness of the restriction might help a reader avoid costly mistakes.
- **Explains a concept or an idea.** This type of topic is most helpful to a reader. However, creating index entries that describe a concept or an idea is fairly difficult. The difficulty is in describing the whole concept in a few words.

Do Not Index Superfluous Entries

Frequently, superfluous entries are in an index because the person creating the index refers to every occurrence of selected words or phrases in the document. The index is not a concordance, but rather an information retrieval device.

For example, assume that the following sentence appeared in the text being indexed:

> Separate chapters of this manual are devoted to the use of disk and tape storage devices.

In this example, the sentence provides no information about disk or tape storage. Therefore, a reader would gain nothing from these index entries:

> disk storage, 37
> ⋮
> tape storage, 37

Include entries in an index only if they refer a reader directly to useful information.

Avoid Entries That Are Too General

Do not use entries so general that they apply to a global level of information. Such an entry is too broad to inform a reader of the entry's corresponding content in the text.

For example, do not include entries such as "File Manager, creating files in" in a book about File Manager; or "features, of Skype" in a book about Skype. The only entries under the name of the application you are documenting address its use, for example, entries such as "installing" or "quitting" the application.

Include Common Industry Terminology

Terms for common procedures and commands might differ, depending on the technology or company. If you know of a common synonym for a process or command, include an entry to direct a reader to the term used in your book, as shown in these examples:

abort, *See* cancel
delete, *See* cut
search, *See* find

Avoid Using Headings as Index Entries

Using headings as the basis for index entries results in an index that is hard for the reader to scan.

Incorrect	Correct
running the QuickView utility, 42	direct virtual memory access (DVMA), using, 87
using direct virtual memory access (DVMA), 87	hidden file, definition, 28
What is a hidden file?, 28	QuickView utility, running, 42

Consider Including "Commands" as a Primary Entry

An editor can help you decide whether including "commands" as an index entry is useful. If you decide to include "commands" as a primary entry, follow it with subentries for each command appearing in the book.

commands

```
aset
chmod
```

Describing a Topic

Once you have determined that a topic merits an index entry, find one or more ways to describe it to a reader. The descriptions that you create become the subjects of the index entries.

To describe a topic, follow these guidelines:

- Anticipate a reader's needs.
- Select the proper words for the subjects.
- Arrange the words for emphasis.
- Create multiple entries (double-post).
- Group the entries.

Anticipate a Reader's Needs

Readers usually use an index to answer one of the following questions:

- Where can I find information about a certain task, term, or topic?
- Does the information described by an index entry tell me what something is, how it works, or how to use it?
- Which pages can I ignore because they contain information I already know?

By anticipating a reader's needs as you describe topics for your index, you can create an index that helps a reader find information quickly and easily.

Include Only Terms a Reader Is Likely to Look Up

When creating an entry, ask yourself whether you would be likely as a reader to look in the index for that entry. Usually, the main entry is a noun, a noun phrase, or a gerund.

Questionable entries include the following:

- Entries starting with irrelevant terms such as "about," "how," or "why"

  Other irrelevant terms include "limitations," "purpose of," and any obscure terms that readers are unlikely to look up in the index.
- Commands or widgets used only in sample programs or examples

  For example, in a book in which an exercise involves creating the buttons "Hello!" and "Adios!", do not include "Hello!" and "Adios!" as index entries.
- Titles of books in a "Recommended Reading" section
- Entries that use the same phrasing as section headings

Select Appropriate Words for Subjects

The words that you select are an abstract of the topic. Choose words that are as descriptive as possible. Using words from the text as subjects of index entries might satisfy readers who know the terminology used in the document. For other readers, provide subjects or cross-references worded so that they can find the desired information without specific prior knowledge.

Use Gerunds

Use a gerund, not an infinitive or a plain verb, as the main subject of an index entry when appropriate. For example, use "initializing" rather than "to initialize" or "initialize."

Identify the Entry Type

Identify each entry for a computing term by including the type of entry after its name. For example, this type might be a command, file, function, or attribute:

`apropos` command

  width attribute

Arrange Words for Emphasis

Typically, make the most important word the first word of the subject. The choice of most important word depends on what you want to stress or what is most important to a reader. For example, if the words you choose to describe a topic are "pixwin background color," the primary entries might be as follows:

background color, pixwin
color, pixwin background
pixwin background color

Use Plural for Main Entries

In general, index the plural forms of nouns.

clients
⋮
servers

The exceptions to this rule are nouns that are only singular or nouns that are only used in the singular in the context being documented. Another exception is proper nouns such as the names of applications.

`get` command
⋮
MySQL database

Assign the Proper Font to the Entry

Keep in mind that certain terms require different fonts if you refer to the terms in different contexts. You also need to include a word such as "command," "file," or "directory" after such terms to further clarify the entry.

`quit` command
quit, in contrast to exit

Group Entries

Grouping entries means combining entries that have common first words into primary entries and secondary entries.

Ungrouped Entries	Grouped Entries
SBus, introduction	SBus
SBus block diagram	block diagram
SBus specifications	introduction
	specifications

In the grouped entries, the term "Sbus" is the primary entry, and the terms "block diagram," "introduction," and "specifications" are the secondary entries.

When selecting a subject to be followed by secondary entries, group subjects properly. Do not merely select a word or phrase for a subject because it is common to several entries. For example, assume that the following subjects appeared in your document:

igneous rock
metamorphic rock
rock music

To isolate "rock" and create three secondary entries is incorrect. If you analyze the use of "rock" in each entry, you can see that it is used in two different ways.

Incorrect	Correct
rock	rocks
igneous	igneous
metamorphic	metamorphic
music	
	rock music

Create Index Entries for Cautions, Notes, and Tips

You want readers to be able to locate the various restrictions or ease-of-use features in your document, including the following:

- Sets of rules
- Value limits (maximums and minimums)
- Incompatibilities between features and options
- Caution notices
- Ease-of-use Notes or Tips

Most of the topics that qualify as restrictions are not stated as such in the text. You must analyze the text to find the restrictions to index. Write index entries so that the subject describes the nature of the restriction.

You can also index those Notes or Tips that provide ease-of-use information or features.

Make sure to enable readers who are familiar with the document organization to ignore the entry if they already know the restriction or feature.

Incorrect	Correct
symbolic names, restriction	symbolic names, maximum length of
Window menu shortcut	Window menu shortcut, F9 key

In many documents, certain restrictions are identified specifically because of their importance. For example, assume that the following Caution notice appears in your document:

Caution – Never turn the system unit on or off while a disc is in the disk drive. You might damage the disc.

Index this restriction flagged with the word "Caution." The entry might be as follows:

turning system on or off, Caution notice

Create Index Entries for Acronyms and Abbreviations

Include an acronym or abbreviation in your index if it is unique to your document or documentation set and if it is not likely to be found in common usage. Many acronyms and abbreviations need not be included in an index. For example, the abbreviations for most units of measure, such as Btu, in., or lb, are not good candidates for indexing.

Alphabetize acronyms and abbreviations as words, rather than as the spelled-out version of the acronym or abbreviation. When you include an acronym or abbreviation in an index, follow it with the word or words from which it was formed. Place those words within parentheses.

CCP (console command processor), 1-5

Double-post the entry by adding an entry for the words that form the acronym or abbreviation, followed by the acronym or abbreviation in parentheses.

console command processor (CCP), 1-5

Double-Posting Index Entries

Double-posting means identifying a topic in two different places in an index. For example, a topic that appears as "address switch" and "switch, address" is double-posted in the index. A topic that appears in three places is triple-posted, and so on.

Entry	Double-Posted or Triple-Posted Entries
power indicator	power indicator indicator, power
C shell command interpreter	C shell command interpreter command interpreter, C shell interpreter, C shell command

Be careful of over-indexing with double-posting. Certain entries do not deserve double-posting. For instance, the following example might be acceptable in a document that refers to only a few commands.

Entry	Double-Posted Entries
`grep` command	`grep` command command, `grep`

However, for a manual with many commands, the entries under "command" might grow too numerous. In this case, rather than creating a primary entry of "command" with many secondary entries, index the commands under the command name and do not double-post the entries. Instead, include a cross-reference.

Entries	Double-Posted Entries
`cat` command `grep` command `history` command ⋮	commands, *See* specific command names

Double-posting increases the number of index entries available to a reader, which broadens the scope of the index. The knowledgeable reader is not forced to scan the index for a general entry when seeking a specific topic.

Double-posting has a dramatic effect on usability. It is an essential technique for creating a high-quality index. Try to double-post entries for all key concepts and important terminology.

Keep in mind, however, that double-posting an index can affect your schedule. Because an extensively double-posted index provides a denser, more comprehensive view of a document's topics, be sure to include enough time in your schedule for double-posting your index.

Creating "See" and "See Also" References

You cross-reference index entries by creating "See" and "See also" references.

When to Use "See" and "See Also" References

- Use a "See" reference when you have so many secondary entries that repeating them is unreasonable.

  configuration, *See* measurement configuration

  ⋮

  measurement configuration

    applying storage thresholds
    calculating line speeds
    defining data fields
    defining entities

- Use a "See" reference to send readers from a broad category to a more specific category.

  The next example is valid only if there are several secondary entries under "display thresholds," "exception thresholds," and "storage thresholds." Otherwise, you would double-post.

  thresholds, *See* display thresholds; exception thresholds; storage thresholds

- Use a "See" reference to direct a reader from a term not used in the document to a term that is used as an index entry.

  cars, *See* automobiles

- Consider using a "See also" reference to direct a reader to related information at another index entry.

  Depending on how your index is structured, you might also use a "See also" cross-reference from a specific category to a general one.

  dBASE, 37

    See also database applications

- Use a "See also" reference to avoid fourth-level entries.

  performance database

  See also update, performance database
  backing up data in
  deleting
  updating
  ⋮

  update, performance database

  automatic
  displaying status
  manual
  starting
  stopping

- Clarify index entries for some noun modifiers, such as "data" and "file," that are ubiquitous.

  If you have several long and complicated entries that start with the same word, readers might not look far enough to find a given topic. In this case, use a "See also" reference to help a reader.

  data

  See also data files; data records
  collecting
  deleting

- Never use a "See" reference with an entry that includes a page number.

  Never include a page number in a "See also" reference.

Incorrect	Correct
structured files, 7-3 *See* files, structured	structured files, *See* files, structured
structured files, 7-3 *See also* chaotic files, 8-4	structured files, 7-3 *See also* chaotic files

- Make sure that a "See" or "See also" reference repeats the exact wording of the entry to which it refers.

Incorrect	Correct
database, *See* PDB	database, *See* performance database (PDB)
⋮	⋮
performance database (PDB)	performance database (PDB)

- Do not use unnecessary "See" references.

  If you can reasonably double-post an entry, do so. Readers have every right to be annoyed if you send them searching through an index just for one or two page numbers.

Incorrect	Correct
command objects, *See* objects	command objects, 5-2, 5-8
⋮	⋮
objects, 5-2, 5-8	objects, 5-2, 5-8

In particular, do not send readers from a specific entry to a general entry, under which they must then search for the specific entry. However, be careful that you include general information under the specific entry as well. In the following example, "changing report attributes" must be under the entries for specific reports because a reader might not look at the general entry.

Incorrect	Correct
forecast reports, *See* reporting	forecast reports, 3-67
	changing report attributes, 3-122
⋮	⋮
reporting	reporting
automatic, 2-33, 3-174	automatic, 2-33, 3-174
changing report attributes, 3-122	changing report attributes, 3-122
forecast reports, 3-67	forecast reports, 3-67
predefined reports, 3-71	predefined reports, 3-71

- Do not use a "See also" reference to send a reader to a duplicate (double-posted) entry in an index.

  Incorrect:

  entry-sequenced files, 7-3

  See also files, entry-sequenced

  ⋮

  files, entry-sequenced, 7-3

How to Use "See" and "See Also" References

The following are basic formatting and punctuation rules for "See" and "See also" references:

- Italicize the words "See" and "See also."

  base window, 45

  See also pane

- Never include page numbers with "See" and "See also" references.

  Use those references to direct a reader to another index entry.

  graphics, *See* figures

- Place the "See" reference on the same line as the index entry, separated with a comma.

  search, *See* find

- Place the "See also" reference at the beginning of the entry.

  Place the "See also" reference on a line by itself, and indent the reference from the line above.

  aggregation scheme, 16

  See also summarization, data

- Use a semicolon to separate multiple "See" and "See also" references.

  local area network, 24

  See also Ethernet; standards, networking; wide area network

- For secondary entries, use "See" and "See also" references as follows.

Incorrect	Correct

files
  comparing, 30
  deleting, 24
  initialization, 121
  renaming, 141
files, editing, *See* `vi` editor
files, printing, 35
  See also PostScript documents
	files
	  comparing, 30
	  deleting, 24
	  editing, *See* `vi` editor
	  initialization, 121
	  printing, 35
	    See also PostScript documents
	  renaming, 141

Avoiding Indexing Problems

This section explains some established rules for indexing. If you disregard them, you might confuse or annoy readers. You might also appear incompetent to any reader who is knowledgeable about indexing.

Use a Single-Level Entry for a Single Topic

If you can use the primary entry alone, do so. A primary term with only one secondary term must be on a single text line. If you feel that the primary entry alone is misleading, rewrite it.

Incorrect	Correct
optimization routines use of, 38	optimization routines, 38 or optimization routines, use of, 38

Use an Adjective With a Noun as a Primary Entry

Use adjectives with related nouns to provide enough information for a reader.

Incorrect	Correct

implicit
  `logoff` command, 6-4
  `open` command, 6-1
  `wait` command, 6-6
	implicit commands
	  `logoff`, 6-4
	  `open`, 6-1
	  `wait`, 6-6

This rule applies to noun modifiers, that is, nouns that are being used as adjectives. This rule eliminates awkward, confusing constructions in which the primary entry is a noun relative to some secondary entries and an adjective relative to others.

Incorrect	Correct
data collecting, 81 files, 90 purging, 62 records, 47 ⋮ `wait` command, 29 parameter, 33	data collecting, 81 purging, 62 data files, 90 data records, 47 ⋮ `wait` command, 29 `wait` parameter, 33

Avoid a Primary Entry That Is Too General

Generally, if a primary entry is followed by half a page or so of secondary entries, either the primary entry is too broad or you are over-indexing. For example, in a printer manual, the primary entry "printer" is too broad to be indexed as a term.

If you feel that it is necessary or helpful, use "commands" plus a "See" reference to send a reader to alternate methods of locating a given command.

Incorrect	Correct
commands alias, 14 `at`, 19 batch, 23 ⋮ `ypmatch`, 132 `ypwhich`, 134 `zcat`, 135	(no entry at all, if most of the document describes commands) or commands, *See* individual commands by name or commands, summary of, 18

Do Not Over-Index

For manuals that have a repetitive structure, do not provide so many entries that they get in a reader's way. For example, in a reference manual containing many commands or utilities, you might be tempted to index the subheadings under each command. This often results in over-indexing, as shown in the following table.

Incorrect	Correct
ast command attributes, 42 syntax, 33 examples, 36	ast command, 33, 42, 36
ast_process command attributes, 49 syntax, 53 examples, 56	ast_process command, 49, 53, 56
ast_subvolume command attributes, 50 syntax, 54 examples, 57	ast_subvolume command, 50, 54, 57

Over-indexing also occurs if you create several secondary entries under a primary entry when all entries are on the same page.

Incorrect	Correct
input devices buttons, 167 dials, 167 digitizer, 167 scanner, 167	input devices, 167

Do not provide two adjacent entries that are very similar. However, always double-post acronyms and abbreviations even if it is possible that the entries might be adjacent in the index. The test: If you omit an entry, can a reader still find the right place in the document?

Incorrect	Correct
delete_file command, 41	delete_file command, 41
deleting a file, 41	

Do Not Under-Index

Some kinds of under-indexing are very obvious because they do not provide enough specific information to be useful.

Incorrect	Correct
reports, 31–39, 77	reports exporting, 77 generating, 34 preformatting, 33 specifying format, 36–39 types of, 31–33

Other types of under-indexing are not obvious, except to a reader. It is especially important to index *concepts*, not just the terms that appear in the document.

Incorrect	Correct
`archive` command, 77	`archive` command, 77 ⋮ backing up data to tape, 77 ⋮ tape backups, 77

Alphabetize by Keyword in Subentries

Alphabetize subentries by keyword, not by beginning articles, conjunctions, or prepositions. Reword subentries so that the keyword or key term, rather than an irrelevant introductory word, appears at the beginning of the subentry. If the subentry requires an article, conjunction, or preposition to flow correctly, try to include these additional words at the end of the subentry rather than at the beginning.

Incorrect	Correct
snapshots and mirror components, 38 in replication sets, 30 license for, 15 managing, 45–57 of legacy volumes, 62 of replication sets, 22	snapshots of legacy volumes, 62 license for, 15 managing, 45–57 mirror components and, 38 in replications sets, 30 of replication sets, 22
user accounts and roles, 82 logging in to, 109 passwords for, 84	user accounts logging in to, 109 passwords for, 84 roles and, 82

Alphabetize by First Letter After a Symbol

For path, file, or variable entries that begin with a symbol, alphabetize these entries by the first letter of the first word following the symbol.

Incorrect	Correct
Symbols	**C**
`_config` file, 9	`cancel` command, 33
`/etc/uucp/Limits` directory, 12	`_config` file, 9
`.info` file, 21	Create menu, 7
`$PATH` environment variable, 8	**E**
E	error reporting, 18–21
error reporting, 18–21	`/etc/uucp/Limits` directory, 12
external files, 24–25	external files, 24–25
	I
	ID numbers, 39
	`.info` file, 21
	P
	`$PATH` environment variable, 8
	primary numbers, 4

Refining and Checking an Index

While creating the first draft of an index, you probably concentrated on the individual entries and their secondary entries. While editing the index, you are concerned with the index as a whole.

Editing an index might require that you create or delete entries, combine or split entries, and regroup or reword entries. In a sense, editing an index is not very different from editing the document. Namely, you verify that all necessary material is included, that it is in the intended order, and that it is error free.

Remember that you are reading an index in an abnormal way, that is, you are reading it from start to finish. Normally, readers search directly for the word or phrase they hope to find. Because you are reading an index this way, you might think that many entries are not necessary or are redundant. For some entries, this might be true, but to delete many entries on that basis alone is risky. Unless your analysis of the topic was incorrect when you created the entry, you probably had a specific reason for adding the entry.

The following sections review common problem areas in indexing.

Spelling

Many publishing systems do not check the spelling in the embedded index entries when the spelling checker is run on the body of the document. For this reason, always carefully proofread and run the spelling checker on index entries to catch spelling and typographical errors.

Differences in Wording

Determine whether slight variations in wording are intentional or whether you should use only one wording. If there are valid subjects that differ only slightly in wording, examine them to be sure that readers can recognize the difference. You might have to reword the subjects to make the differences apparent.

After creating the index, you might discover that you have used inconsistent terminology in the document. A consistency check and any necessary corrections are well worth the time it takes to standardize terminology.

For example, "AdminTool" might also appear as "Administration Tool"; "Admin Tool"; and "setting up, configuring." You also must standardize usage for terms such as "home directory" and "root directory," and "superuser" and "root user."

Misused Singular Forms and Plural Forms

Check the entries for the misuse of singular forms and plural forms. Usually, only one form of a subject is justified. If you find more than one, combine secondary entries under one subject. Using both forms of a subject, such as "data set" and "data sets," can cause errors. Several other entries and their subsequent secondary entries might intervene between the singular and plural forms of a subject. Spare readers the trouble of checking the index for both forms.

Incorrect	Correct
data set	data sets
input, 31	address, 23
output, 33	area, 27
data set address, 23	format of, 19
data set area, 27	input, 31
data sets	output, 33
format of, 19	table of, 40
table of, 40	

Effective Double-Posting

Check that all meaningful variations of a subject's wording appear in an index. See "Double-Posting Index Entries" on page 311.

Number of Page References for Entries

Include no more than two to four page references per index entry. If an entry has more than two to four page references, see whether you can create secondary and tertiary entries to reduce the number of page references. See "Online Index" on page 329 for more information about providing separate entries for the benefit of online indexes.

Incorrect	Correct

block diagram, 21, 28, 33, 37, 45	block diagram
	    attribute generator, 33
	    frame buffer, 37
	    front-end processor, 28
	    SBus adapter, 21
	    system unit, 45

Proper Topic Cross-References

Page references for each occurrence of a topic should be the same and should appear in each place. In the example, a reader looking up "operator messages" would not be aware of all the other places where information exists. Create secondary entries under "operator messages" and give readers the same information they would have found had they looked up "messages."

Incorrect	Correct

messages
    from operator, 2-34
    to operator, 2-15, 3-7
    to programmer, 5-12
    ⋮
	messages
	    from operator, 2-34
	    to operator, 2-15, 3-7
	    to programmer, 5-12
	    ⋮

operator messages, 2-34, 3-7
⋮
	operator messages
	    from operator, 2-34
	    to operator, 2-15, 3-7
	    to programmer, 5-12

"See" and "See Also" References

If a "See" target has only one or two page references, repeat those page references at the originating entry. One exception is if you are trying to direct the reader to terminology used in the book or product. Read "Creating "See" and "See Also" References" on page 313.

Check the wording of the subjects in "See" and "See also" references. During the creation or regrouping of entries, the wording of the subject might have changed. Make sure that the target entries are still in the form used in the "See" and "See also" reference.

Bad Page Breaks and Bad Column Breaks

One form of a bad page break results when a primary entry with multiple secondary entries and perhaps tertiary entries breaks in the middle at the foot of the last column on a right page. The first column on the following page begins with an indented secondary or tertiary entry. Bad page breaks cause problems for a reader, who must look back to the previous page to find the primary entry.

Correct bad page breaks by repeating the primary entry above the carried-over secondary entry followed by the word "continued" in italic and surrounded by parentheses.

graphical user interface
  menus
    Edit, 8
    File, 10
    Format, 12
    general navigation, 3

graphical user interface
  menus (*continued*)
    Graphics, 16
    Special, 20
    Table, 21
    View, 22

Avoid having a single primary entry at the beginning of an alphabetic section at the bottom of a column. Force the alphabetic character to the top of the next column, carrying the single primary entry along with it.

Likewise, do not leave a single line at the end of an alphabetic section at the top of a column. Force a column break one or two lines before the widowed line.

Note – Some authoring tools do not allow you to control page breaks and column breaks in indexes.

Secondary Entries

When refining your index, examine secondary entries carefully in the following problem areas.

Levels of Secondary Entries

Check the levels of secondary entries so that proper indentation shows the relationship of one entry to the preceding one. See "Nested Index Entries" on page 302.

Redundant Secondary Entries

Main entries that are followed by secondary entries should not have the same page reference.

In many cases, you can eliminate the secondary entries because a reader can find all the information on one page. Redundant secondary entries often occur when items in a table are indexed. See "Do Not Over-Index" on page 318.

Incorrect	Correct

data sets
    format of, 2-7
    table of, 2-7
	data sets, 2-7

Possible Primary Entries in Secondary Entries

Check whether a secondary entry can also appear as a primary entry. If it can, verify that it exists as such or create the primary entry and insert it in the proper place.

Possible Rearrangement

Check whether secondary entries can be rearranged to stress a certain point. In this example, all three secondary entries can be in the same form. The form depends on what you want to stress.

Incorrect	Correct

window system
    colors, changing, 8-11
    icon, moving, 4-22
    saving properties, 8-15
	window system
	    colors, changing, 8-11
	    icon, moving, 4-22
	    properties, saving, 8-15

Appropriately Combined Secondary Entries

Review an index to make sure that you have combined relevant entries.

Incorrect	Correct

PlirgPak
    selection protocol, 2-4
 	
PlirgPak atoms, 4-8
 	
PlirgPak drag-and-drop atoms, 4-4
 	
PlirgPak drag and drop handshaking, 4-2
 	
PlirgPak integration
    why do it, 1-2
	PlirgPak
	    atoms, 4-8
	
	    drag and drop
	
	        atoms, 4-4
	        handshaking, 4-2
	
	    integration, 1-2
	
	    selection protocol, 2-4

Secondary Entries Under More Than One Topic

Check for secondary entries that can be arranged under one topic rather than several. Such division of secondary entries is usually the result of misused "See also" references.

Incorrect	Correct

find function
    See also search function
    examples, 8-22
    use of, 8-15
    variables, 8-18
    ⋮
 	
search function
    dialog box, 8-9
    and replace function, 8-21
	find function
	    dialog box, 8-9
	    examples, 8-22
	    and replace function, 8-21
	    use of, 8-15
	    variables, 8-18
	    ⋮
	
	search function, *See* find function

Secondary Entries for Combined Terms

If similar terms appear in separate combined terms, move secondary entries under the appropriate combined term.

In the next example, the secondary entries starting with "database" under the "classing engine" main entry belong under the "classing engine database" main entry.

Incorrect	Correct

classing engine
    adding a new file type, 6-6
    attributes, 6-4
    database, accessing, 6-7
    database, converting, 6-8
    database, reading, 6-8
    interactive modification, 6-5
    mapping function, 6-3
classing engine database
    location of, 6-7
    network, 6-9
	classing engine
	    adding a new file type, 6-6
	    attributes, 6-4
	    interactive modification, 6-5
	    mapping function, 6-3
	classing engine database
	    accessing, 6-7
	    converting, 6-8
	    location of, 6-7
	    network, 6-9
	    reading, 6-8

Secondary Entries Under Various Forms of One Topic

The number of secondary entries under various forms of the same topic should all be the same. In the example, "attention key" appears after the "terminal, communications" entry so that readers are aware of the information regardless of how they look it up.

Incorrect	Correct

communications terminal
    attention key, 4-16
    polling character, 4-11
    READY indicator, 4-10
    ⋮
terminal, communications
    polling character, 4-11
    READY indicator, 4-10
	communications terminal
	    attention key, 4-16
	    polling character, 4-11
	    READY indicator, 4-10
	    ⋮
	terminal, communications
	    attention key, 4-16
	    polling character, 4-11
	    READY indicator, 4-10

Checking the Size of an Index

After you have edited the index, compare the size of the index with the size of the document. Although the index size is not an indication of its quality, an index that is too small for the size of the document is suspicious and might indicate serious omissions.

A minimum length for an index is one page of index entries for every 20 pages of text, or about one index entry for every 100 words of text. This can be considered a "5 percent" index. For dense technical material, however, this guideline is too low. For a dense technical manual, try to provide one page of index entries for every 10 pages of text, which can be considered a "10 percent" index.

If you check the length of your index by page count rather than by word count, do not count text pages that contain any of the following if they occupy more than about two-thirds of a page:

- Flow diagrams
- Figures
- Code examples
- Front matter (title page, table of contents, and so on)
- Blank space longer than three-quarters of a page
- Glossary or bibliography

If the index falls below the guidelines, check that all topics in the document are entered in the index.

Global Index

The type of index most writers work with is the "back of the book" variety. A *global index* combines the back-of-the-book indexes from all the books in a set. A global index is a valuable information retrieval device for a reader who is not familiar with all the books in a set. A global index provides a single place where readers can find the information they seek without having to look through several individual indexes.

Formatting a Global Index

In a global index, merely referring a reader to page numbers is insufficient. Because a global index combines the indexes from several books in a set, a reader also needs to know the book in the set to which the reference applies. For this reason, a global index requires a special page numbering style. One method is to use a four-letter abbreviation for an entry's book title, with a running footer that provides a legend on each index page for the abbreviations. If such a footer is not possible or desirable, put a legend for abbreviations at the beginning of the index.

Editing a Global Index

Because of its size and complexity, a global index is by far the most difficult type of index to create properly. You cannot just combine the indexes from several books and assume that, if the previous back-of-the-book indexes were correct, a global index will also be correct.

When combining indexes to make a global index, the indexer must edit the global index for most of the indexing mistakes described under "Refining and Checking an Index" on page 321. Specifically, the indexer must check for the following:

- Consistent page numbering style
- Levels of secondary entries
- Differences in wording
- Misuse of singular and plural forms
- Meaningful variations (double-posting) of a subject's wording
- Possible rearrangement of secondary entries
- Bad page and column breaks

Note – Some authoring tools do not allow you to control page breaks and column breaks in indexes.

Many of these mistakes might be in a global index even though they might not be in the original indexes. These mistakes happen when common terms in different indexes are combined, slightly different terminology is used in different books, and different indexing choices have been applied in each book.

You can fix the problems that result from creating a global index in one of two ways:

- Edit the index entries in the original text and re-create the index.
- Edit the global index and leave the index entries as they are in the original text.

Online Index

If your writing tool supports an online index, consider including one to help readers scan the content of your document and quickly find what they need.

Multiple page numbers translate into many links next to an entry in an online index. So while in print you can collapse repetitive entries, for online indexes elaborate them as much as possible to help the readers find the correct entry quickly.

In the following example, the page icons in the online index entries are used to replace page numbers, which do not apply online.

Print Index	Online Index
ast command, 33, 42	ast command
ast_process command, 44, 49	attributes,
	syntax,
	ast_process command
	attributes,
	syntax,

An online index simplifies and expedites a reader's search for specific information much more than a search engine. A search engine alone can be inadequate for finding information quickly, especially if the tool searches a large database of documents.

For instance, a search engine scans text looking for occurrences of the word or phrase typed in the search box. Then, it lists every document that contains any mention of the word or phrase. A reader can lose a considerable amount of time sifting through these documents before finding the desired information. Also, a search engine cannot provide a topic analysis or overview of the document.

In addition to following the guidelines for a good print index, a good online index makes full use of hypertext by linking index entries to the relevant text in your document.

APPENDIX A

Developing a Publications Department

As a member of a publications department, your charter is to develop, write, edit, validate, and publish documentation that supports the information needs of your customers.

This appendix provides guidance to help you meet the goals of that charter by giving you information on aspects of publications departments ranging from staffing concerns to technical review procedures. It is primarily intended for companies undergoing rapid growth in their documentation requirements. While most publications-related issues are covered, this appendix does not deal with general management issues such as hiring or personnel reviews.

For a list of books that cover the topic of management issues in more detail than can be provided in this appendix, see "Project Management" on page 426.

Note – This appendix explains as an example how the publications department fits into a software computer company's organization. You might have to modify some of the recommended processes and procedures to match your own situation.

Topics discussed include:

Establishing a Publications Department

The documentation presence in a company usually begins with a few people producing written material to accompany products. Publications departments can range from a single permanent publications manager working solely with outside contractors to a department featuring several writing, illustration, and production groups. Table A–1 describes the maturity levels of documentation organizations and their goals at each level.[1]

Use this table to see where your department fits on the continuum. When analyzing your organization, judge its performance as a whole over a long period of time. Although your organization might exhibit some features of a particular level, one instance does not define your organization as having achieved that level of maturity.

TABLE A–1 Process Maturity Levels of Documentation Organizations

Level	Description	Publications Project Manager	Transition to the Next Level
Level 0: Oblivious	Unaware of the need for professionally produced publications. Publications are produced by anyone who is available and has time.	None	Staffing with professional technical communicators
Level 1: Ad hoc	Technical communicators act independently to produce publications with little or no coordination. They may be assigned to different technical managers.	None	Development of a style guide
Level 2: Rudimentary	The beginning pieces of a process are going into place. Some coordination occurs among the technical communicators to assure consistency, but enforcement is not strong.	None to very little	Introduction of some project planning

[1] Table and description from JoAnn T. Hackos, *Managing Your Documentation Projects* (New York: John Wiley & Sons, 1994), p. 47–48. Used with permission.

TABLE A–1 Process Maturity Levels of Documentation Organizations (Continued)

Level	Description	Publications Project Manager	Transition to the Next Level
Level 3: Organized and repeatable	A sound development process is in place and being refined. People are being trained in the process. Project management is in the beginning stages, with senior technical communicators learning the rudiments of estimating and tracking.	Introduction of project management	Strong implementation of project management
Level 4: Managed and sustainable	Strong project management is in place to ensure that the publications-development process works. Estimating and tracking of projects are thorough, and controls are in place to keep projects within budgets and schedules. Innovation gains importance within the strong existing structure.	Strong commitment to project management	Beginning of the implementation of more effective processes
Level 5: Optimizing	Everyone on the teams is engaged in monitoring and controlling projects. As a result, effective self-managed teams are becoming the norm. Innovations in the development process are regularly investigated and the teams have a strong commitment to continuous process improvement	Strong commitment to project management and institution of self-managed teams	Strong and sustainable commitment to continuous process improvement

Establishing the Value of the Department

If you want to expand your department, the first step is usually to convince management of its value. However, measuring the "value added" of accurate, comprehensive documentation written by professional technical communicators is not easy. Your focus is on added ease-of-use for the user, in the product interface as well as the documentation. However, that is often not as important an argument to management as actual costs saved.

This section provides some ways in which you can show how good documentation adds value, and offers some metrics from other studies that you can use.[2]

Accounting for Value Added

Traditional accounting practices often make showing the benefits of improving quality very difficult. For example, many accounting systems still track costs by department rather than project. If customer support costs go down, the customer support group looks good. The documentation group does not get any credit for reducing support costs, even if good documentation contributed substantially to the reduction.

Many accounting systems are still based on a manufacturing model rather than a labor-intensive service model. For example, a documentation group that is measured only on pages per day appears to cost more if their higher-quality documents have fewer pages. Value that the documentation group is adding through activities other than writing and production (such as interface evaluation) or the greater benefits of shorter documents might not be reflected in the accounting reports.

Tracking Avoided Costs and Costs Saved

When considering the value added by a professional publications staff and good documentation, costs avoided are as significant as costs saved. For example, suppose a company gets 100,000 support calls a year at a cost of $30 a call. If better documentation reduces the call volume by 10 percent (either the number of calls or a shorter duration of call), the technical communicators save the company $300,000 a year.

Another aspect you can point out is the cost per problem at different points in the product development cycle. Problems found in the writing and editing cycle are much less expensive to fix than problems found once the product is in the field.

Measures that show *increased benefits* resulting from good documentation include:

- More sales
- Increased productivity
- A higher percentage of forms or response cards returned
- Forms or response cards returned more quickly
- More users' problems identified early in the process

Measures that show *reduced costs* resulting from good documentation include:

- Fewer support calls, which result in lower support costs
- Less need for training, which results in lower training costs
- Fewer requests for maintenance, which results in lower maintenance costs
- Less time needed for translation, which results in lower translation costs

[2] The material in this section is based on information from Janice C. Redish, "Adding Value as a Professional Technical Communicator," *Technical Communication* 42:1 (1995), p. 26–39. Used with permission.

- Less effort (time, lines of code, rework) is needed when technical communicators are involved early in the development process than when they are not
- Lower costs for writing, paper, printing, and so on because technical communicators showed developers that they did not need all the documentation that they were planning
- Fewer errors in specifications written by technical communicators than in specifications written by engineers

When trying to find ways to illustrate *value added* by good documentation, consider the following techniques:

- Estimating avoidable costs from historical data, for example, costs of writing and sending updates and bulletins about problems and solutions, support costs, and costs to the customer's company.
- Comparing two documents or two situations where one document used the services of publications personnel and the other document did not. For example, you might submit a prior version of a document that has been revised, or interview users who have access to the documentation and those users who do not have access.

Establishing Expertise

Companies, especially those with fledgling publications groups, often regard documentation as merely writing down what the product does. You should also try to establish a role as the user advocate. You can offer your expertise during the project development stage in areas such as interface evaluation, menu item and error message wording, usability, and so on. If you encounter resistance to early writer involvement, point out that problems like inconsistent interface features take much less time, and therefore money, to correct at early stages in product development than if the writers are not involved until the actual writing cycle begins.

Funding the Publications Department

Once you've convinced management that your department should be expanded, you might be asked to help determine how your publications department is funded.

Some possibilities for funding are as follows:

- Funding by the division to which the publications department reports regardless of who receives the documentation services. For example, a publications department might report to and be funded by the marketing division even though the department is developing documentation for the engineering group.
- Funding for department personnel from the budget of the project they are documenting. The publications manager identifies the staffing level required for a given project and the positions are funded by the project's budget.
- Funding for centralized services such as editing, illustration, and production by the division to which the publications department reports, with individual writers funded by the project they are documenting.

Obviously, each scenario has advantages and disadvantages. For example, if you are a writer reporting directly to an engineering project manager, you will probably have a closer relationship with the engineers on the project than if you are a member of a separate publications department. At the same time, your concerns as a publications-oriented team member might not be taken as seriously in an engineering group as they would if you reported to a publications manager.

Determining the Roles of the Publications Team

When you have permission to expand your staff, consider the roles your staff members need to fill. The following sections describe the roles of various publications personnel.

Manager

- Supervises all personnel matters: recruiting, hiring, training, supervising, and evaluating
- Plans, schedules, and might oversee projects
- Acquires and allocates resources: monetary, personnel, equipment, outside resources
- Works with project initiation, design, and planning teams

Writing Team Leader

- Coordinates and maintains the documentation plan (see "Writing a Documentation Plan" on page 341 for more details)
- Coordinates and tracks schedules
- Assigns writing tasks to individual writers and consults with them to set priorities
- Makes sure technical or hardware problems encountered by writers are resolved
- Represents the writing team at project team meetings or other department meetings relating to the project
- Holds weekly meetings of the writing team and issues weekly status reports
- Coordinates multiple technical reviews
- Serves as the liaison with the production staff

Writers

- Determines scope and contents of assigned books with input from other writing team members, marketing, usability testing, and other relevant departments
- Writes the documentation plan if there is no writing team leader (see "Writing a Documentation Plan" on page 341 for more details)
- Gathers and verifies source data, which includes attending engineering and product design meetings

- Disseminates project information to the rest of the documentation team
- Writes the documentation according to the audience and technical content defined in the documentation plan
- Develops or works with an illustrator to create illustrations
- Incorporates editorial and technical review comments
- Arranges for validity and usability testing

Editor

- Directs and guides the writers to write clearly and consistently, with the user in mind at all times
- Provides editorial support at all levels: developmental editing, copy editing, and proofreading (see "Types of Editing" on page 218)
- Ensures that grammar, syntax, and spelling are correct in all documents
- In projects with more than one writer, brings the different styles of the various writers into a consistent whole, to achieve a single voice
- Maintains the project style sheet
- Keeps all writers on a project informed of stylistic decisions, title changes, and so on
- Ensures correct use of copyright and trademark information
- Checks that all figures and tables (from chapter to chapter and book to book) are consistent in style and quality, are in the right position, and, if appropriate, are numbered correctly

Graphic Designer

- Develops the overall look and design of the documentation product
- Determines how typography, use of color, and general layout of the information are handled
- Designs graphic elements that assist a reader, such as icons and glyphs.
- Possibly produces graphics for the software interface or online documentation
- Designs the documentation format for a company, a set of documents, or a single document
- Designs packaging and manual covers

Illustrator

- Works closely with the writer to create drawings for information products
- Participates in the project documentation team and understands the material, especially on large projects
- Works with the graphic designer on projects that require numerous illustrations

Deciding on Contract or Permanent Staff

Once you have determined who is paying for the publications staff and defined their tasks, you can start hiring staff. One of the primary decisions in this area is whether to hire permanent staff or to use contractors. Small companies might choose to hire a documentation manager who supervises a staff of contractors. Larger companies might have a mixed staff of contractors and permanent writers, or they might hire contractors for peak loads or short-term projects only.

Advantages of Using Contractors

- **Expertise.** Full-time contractors can offer a high level of experience and professionalism. These qualities are especially important if your publications department is young and your processes are not fully in place.
- **Flexibility.** In slow times, you do not need to maintain a full staff.
- **Opportunity to see prospective employees in real work conditions.** More and more companies are using contractors with an option to convert them to permanent staff. This practice enables you to see whether the employee is productive and works well in your environment.
- **Cost.** The cost per hour of contractors is necessarily higher than permanent staff. However, you save on paying for health benefits and vacation, and sometimes the office space and equipment costs, of a permanent position. Also, experienced contractors can often produce documentation quickly.

Disadvantages of Using Contractors

- **Learning curve.** Contractors are usually unfamiliar with your product, your processes, and the employees of your company, whether in your own department or in other departments from which they need to gain information.

  Not only must you allow time for the contractor to "get up to speed," but the knowledge gained also departs with the contractor rather than benefiting the company.

  Also, for editors especially, lack of familiarity with your in-house style and the lack of an established relationship with in-house writers can be difficult factors to overcome.
- **Accessibility.** If contractors are working off site, they are not available for spur-of-the-moment meetings or decision-making.
- **Communication.** The adage "out of sight, out of mind" is unfortunately often true when you mix permanent staff and contractors. For example, sometimes tacit agreements on style or processes are made in casual hallway conversations that the contractor misses and that the on-site personnel forget to pass on.

Considerations When Hiring Contractors

After deciding to hire a contractor, make sure that you consider the following factors:

- **On-site time commitment.** Establish a clear understanding of how much time the contractor is expected to be on site, for meetings or other necessary commitments.
- **Tax regulations.** Make sure that you comply with the regulations of federal, state, and local tax agencies. Some payment agreements and on-site time considerations might affect whether tax agencies will consider contractors as permanent employees.
- **Nondisclosure agreement.** You should have the contractor sign a standard agreement that protects your company's confidential or proprietary information.
- **Compatible software delivery mechanism.** Make sure that the contractor has compatible software and hardware and can deliver easily into your current system.
- **Delivery of material.** Materials such as updated schedules and project style sheets must be delivered to the contractor. Material to be received from the contractor includes drafts or schedule updates. You should have an easy and efficient method of delivery.

Considerations When Hiring Permanent Staff

If your company has a personnel office, hiring procedures are probably already established. However, if your hiring is less formal, you might need to consider the following tasks:

- Developing job descriptions
- Determining grade or salary levels for levels of writers, editors, or illustrators
- Evaluating the level of seniority needed for a position (for example, could you use entry-level applicants or college interns?)
- Establishing the interviewing team (for example, should developers on the project be included?)
- Conducting orientation and training

Scheduling

One of the more difficult tasks facing any publications department is developing accurate and realistic schedules, and modifying them while the project is underway. This section provides some basic information about schedule estimates and contingencies, but it is by no means exhaustive. For a list of books that deal specifically with managing documentation projects, see "Project Management" on page 426.

Estimating Task Times

The following table shows a rough formula for calculating the hours needed for documentation tasks. Keep in mind that these are *estimates only* and that they might vary depending upon the nature of the documentation. For example, very technical documentation is usually more time-consuming to write and edit than overview information. Outside factors such as poor source material or limited availability of subject matter experts can also affect schedules.

TABLE A–2 Productivity Estimates

Activity	Formula for Calculating Hours
Writing new text	3–5 hours per page
Revising existing text	1–3 hours per page
Editing	6–8 hours per page
Indexing	5 pages per hour
Production preparation	5 percent of all other activities
Project management	10–15 percent of all other activities

You might want to consider setting up a system to track the amount of time your staff members spend on each project. You can use this data to more accurately predict the amount of time needed for future projects.

Developing a Publications Schedule

When developing the schedule, tie your deliverables to project milestones rather than calendar dates. Estimate the time before or after a milestone at which you expect to deliver the component (for example, "The first draft of the documentation will be completed two weeks after the alpha version of the software is delivered to Product Test"). That way, you can more easily adjust your documentation schedule to match the progress of the project. Be realistic in your own assessment of actual progress in other departments, such as product development and testing.

The following table shows a typical publications project schedule.

TABLE A–3 Sample Publications Schedule

Milestone	Date Information
Engineering specification	Date from engineering
Documentation plan	Start and end dates
Alpha software delivery	Due date
First draft	Due date
Technical review	Start and end dates

TABLE A–3 Sample Publications Schedule *(Continued)*

Milestone	Date Information
Developmental edit	Same start and end dates as technical review
Usability test of draft	Same start and end dates as technical review
Index development	Same start and end dates as technical review
User interface freeze	Date from engineering
Illustrations complete	Due date
Feature/function freeze	Date from engineering
Second draft	Due date
Copy edit	Start and end dates
Validity testing	Same start and end dates as copy edit
Final draft	Due date
Proofread	Start and end dates
Final draft to production (hard-copy and online versions)	Due date

Be sure to keep track of changes in delivery dates, and let others involved in the project know whether a date is going to slip. Setting expectations up front is the best way to establish credibility and to call attention to late deliverables on which your own deliverables depend.

Documentation Process

This section walks you through the process of planning documentation, writing it, and getting it reviewed.

Writing a Documentation Plan

The documentation plan informs the rest of the product team about your plans for the product documentation. The plan should be reviewed by representatives of all departments involved with the product. Keep the plan up to date throughout the project, with major changes being announced when necessary.

The documentation plan is based on input from various departments. The following table provides some ideas of the types of information you might get from other departments.

TABLE A–4 Documentation Plan Input From Other Departments

Department Name	Relevant Information
Marketing	Product definition, product name, customer profile, feature and function product requirements
Development	Feature and function schedule, user interface concerns, error message data, names of experts on subject matter

TABLE A–4 Documentation Plan Input From Other Departments *(Continued)*

Department Name	Relevant Information
Legal	Trademarks, product names
Customer Support	Customer profile, previous product troubleshooting logs, ways to obtain technical support, names of experts on subject matter
Manufacturing and Operations	Part numbers, product packaging, production schedules, shipping lead time

The following table lists the components of a typical documentation plan. It is a sample only, and not all sections are relevant to all product types or publications departments.

TABLE A–5 Documentation Plan Sample Topics

Topic	Content
Product information	Product name and version, brief description of the product's intended use.
Documentation resource requirements	Personnel, equipment.
Revision information	Differences from the documentation of previous versions of the product, if any.
Documentation objectives	Overall objectives of the documentation set or of each book.
Documentation overview	Full list of documentation deliverables and the format in which they will be delivered.
Documentation descriptions	Brief description of the chapters and appendixes in each book, with the estimated page counts. If the books in the documentation set are large, they each might require a separate documentation plan.
Documentation schedule	Publications schedule milestones and other milestones from the project schedule that affect the documentation. Publications milestones might include draft delivery dates, technical review dates, and the final documentation delivery date to production.
Technical review	List of technical reviewers and the projected review schedule.
Test plan	Plans for validity testing, usability testing, and, if relevant, media testing plans and dates.
Edit plan	Editing schedule and book priorities.

TABLE A–5 Documentation Plan Sample Topics *(Continued)*

Topic	Content
Localization plan	Languages into which the document will be translated, required resources, and schedule.
Documentation design	Format of the documentation components, including book sizes, online media, and so on.
Production plan	Printing and packaging specifications.
Issues	Any projected issues that might affect documentation, such as suspected schedule slips, engineering uncertainties, or known project design difficulties.
Critical dependencies	Items needed from other groups, including the due date and the impact if the information is not provided or is late. These items might include the following items: ■ Prototype delivered ■ Subject matter experts designated ■ Technical product specification received ■ User interface frozen ■ Features and functions frozen ■ Alpha version of working software available ■ Installation specifications completed ■ Beta software released to testing ■ Technical review sign-off meeting held ■ Development frozen ■ Final list of error messages, causes, and remedies received ■ Product name confirmed

Coordinating With Product Development

Documentation concerns especially affect product development at two points in the schedule: during technical review and at the end of the project.

- **Technical review.** Your subject matter experts *must* dedicate time to complete a thorough and expert review of your documentation. Unfortunately, technical reviews usually occur when development is busy with last-minute engineering and bug fixes. Management support is often crucial in making sure that the development staff take the time to review the documentation thoroughly. Documentation must be considered an important part of the product by all departments for technical review to be treated with the attention it requires.
- **Code freeze.** The other development concern arises at the end of the product cycle, when you need to freeze your screen illustrations and descriptions of the product but the development staff is still fixing bugs. If you provide printed manuals with your product, make sure that the developers, and management, realize that the documentation requires production and printing lead time. They also need to realize that changes to the product after the documentation has gone to production are difficult, if not impossible, to incorporate and will cost more. Also, regardless of whether the documentation is printed or presented online, finding all locations in the document where descriptions or screen illustrations need to be changed at the last minute can be difficult.

Writing Process

This section deals with the writing of the documentation and with the handoffs the writer must deliver. The following figure shows a sample process.

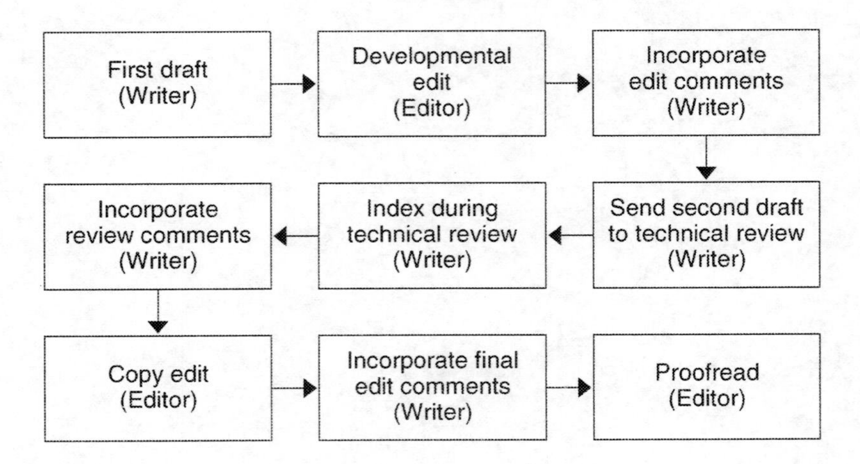

First Draft

The first draft focuses on where information goes and how it should be presented. When writing the first draft, the writer should focus on these tasks:

- Determine organization. Fill out outlines from the documentation plan, making adjustments when necessary. Work with other writers and the editor to define the interrelationships among the manuals.
- Ascertain where technical information is missing or incomplete, and alert engineers and the other writers.
- Develop thematically unified examples (a scenario), if needed.
- Identify terminology issues, and raise them for discussion and resolution.
- Develop an approach to the audience defined in the documentation plan, and develop the voice appropriate for that audience.
- Develop ideas for art and screen illustrations, and compile a preliminary list of figures.

Second Draft

The second draft should be as complete as the state of the software and the specifications permit. It should be fully illustrated. The writing should be polished.

In the writing of the second draft, the writer should focus on these tasks:

- Incorporate edits and comments from reviewers of the first draft.
- Revise for clarity and consistency of voice and terminology.
- Fill in the technical gaps from the first draft as information becomes available.
- Make sure that all terminology and usage decisions made since the first draft are reflected.
- Compile a list of index entries, and begin to develop main entries, subentries, and cross-references.
- Arrange for a validity test (or perform each procedure yourself) for all chapters to ensure technical accuracy and completeness.
- Complete assembly of screen illustrations, and incorporate other types of illustrations.

Editing Process

An ideal editorial cycle includes:

- **Developmental edit** – A review of structure and organization at the first-draft stage
- **Copy edit** – A review of readability and style at the second-draft stage
- **Proofreading** – A final review prior to production

As your company's publications needs become greater and your publications staff begins dealing with more projects, you will probably want to develop a company style guide. This guide should cover editorial and writing guidelines specific to your publications style and your product line. Besides using *Read Me First!* as a model, you might want to examine some of the books mentioned in Appendix E, "Recommended Reading."

For more information about working with an editor, and for checklists describing the different levels of edit, see Chapter 11, "Working With an Editor."

Illustration and Graphic Design

If your illustration and design needs are minimal, someone on your team, such as a writer or publications manager, might be able to handle graphic design and illustration coordination. However, as your needs increase, you should consider adding a production coordinator to your staff. This person can coordinate with outside vendors, contract for illustration and graphic design help, or hire permanent staff in this area when necessary.

Illustration Concerns

Processes for dealing with illustrations vary depending upon the type of illustrations, the availability of compatible screen capture software, your staffing (in-house or contractors), and your budget.

If screen capture software is available for the operating system you are running, the writer is usually responsible for capturing the screen illustrations. The writer is the one most familiar with what the screen should illustrate and with the software being documented and so can more efficiently set up the screen appropriately. However, you might want to send these illustrations on to an illustrator to be "cleaned up" for the best final reproduction quality.

Concept illustrations, on the other hand, usually benefit from having an illustrator render them. Not all writers are gifted artistically, and the time spent producing art is time *not* spent writing. Illustrators are also more familiar with the "language" of illustrations and can usually produce a more professional concept illustration.

The writer, illustrator, and editor should meet periodically to discuss rough sketches and specifications for illustrations.

A preliminary list of illustrations required for the project is produced by all the writers. (For a sample illustration request form and art tracking form, see Appendix D, "Checklists and Forms.") The writers periodically review the illustrations for their specific books with the illustrator until both are satisfied with the final art.

Note – If your documentation consists in large part of updating existing documentation for a standard set of products, consider archiving generic illustrations of your product or its basic concepts for use in future documentation.

Graphic Design Process

The graphic design for your documentation set might already be established or might be standard for each book or online component. However, you might need a graphic design plan if you are adding new types of components to your documentation set, or if you have a new product line for which you want a different look. The following figure illustrates the graphic design process.

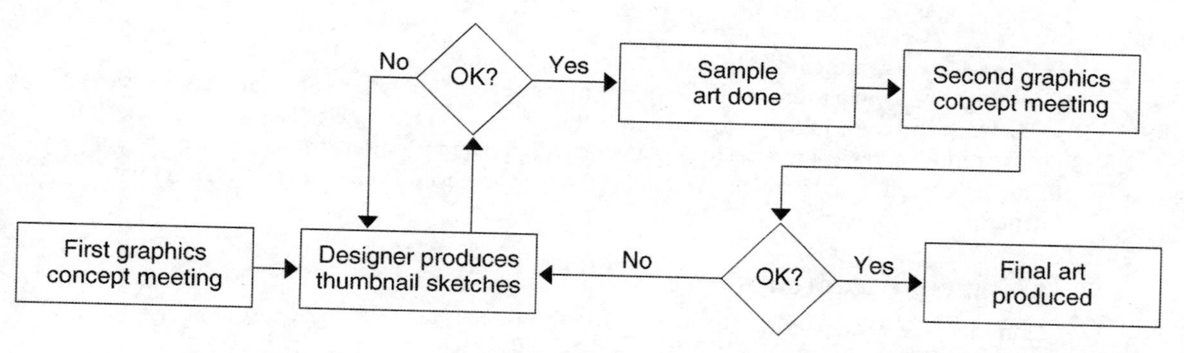

- **First graphics concept meeting.** The writers, editor, and graphic designer should hold an initial meeting to identify and discuss product concepts that could best be conveyed through graphics, page layout, and other documentation design issues. The result of this meeting is a graphic design direction for the project. A graphic designer can often also help in designing online tutorials or other online documentation.
- **Second graphics concept meeting.** A second graphics concept meeting is held to approve the final graphic design. If rework is necessary, the graphic designer either produces another sample for another review or executes the art with the final changes incorporated.

Technical Review

This section explains some of the issues related to the publications team and the reviewers during a technical review process. Remember, thorough technical review is an important contribution to the accuracy and usefulness of your documentation. Unfortunately, this concept is sometimes difficult to convey to other groups. This section provides information on how to promote and conduct a technical review.

Tip – If your company uses wikis internally, consider using a wiki to conduct the review. The wiki format is useful for collecting comments on existing material and incorporating new content from reviewers that clarifies technical information.

Make sure that you provide details about the technical review, including its schedule, in your documentation plan or during planning meetings. Emphasize the importance of having accurate documentation that is thoroughly reviewed and tested. Point out the advantages to the whole company in saved support costs, a good reputation, and favorable industry reviews.

Make sure that the relevant departments allow time for their personnel to review the documentation and that they know the seriousness of the review. For successful completion of a documentation project, the reviewers must be available to conduct reviews and to attend review meetings at the time specified in the documentation plan or the review cover letter. (See "Technical Review Cover Letter" on page 415.)

Once reviewer comments have been returned and evaluated, hold a review meeting. This meeting should help to ensure the accuracy of the documentation, resolve conflicts between reviewers, and collect final comments. (See "Review Meeting" on page 350.) At the conclusion of the technical review and the review meeting, designated reviewers from each department should "sign off" that the documentation, with the agreed-upon changes, meets with their department's approval.

Comment Acceptance

Comments regarding technical *errors* in the documentation should be incorporated.

Comments regarding technical *completeness* should be accepted or rejected depending on the scope and intent of the documentation as specified in the documentation plan or as interpreted by the writer, writing team leader, or manager.

Comments regarding the *purpose and scope* of the manual, including the intended audience, should be accepted or rejected based on the related material in the documentation plan.

Comments on *editorial style*, *organization*, or *cosmetic changes* should be accepted or rejected based on style and design guidelines.

Participants in the Technical Review

Participants in the technical review process should ideally already be members of the project team. However, even those outside the project team, such as customer support personnel and testers, might be able to provide valuable feedback.

Departments you might want to have represented, and their review responsibilities, are listed here. You might want to provide this information, modified for your own situation, as part of the documentation plan.

Product Development

- The documentation accurately describes the technical aspects of the product.
- The text, examples, and illustrations are technically correct.
- All technical information specified in the documentation plan is included.

Product Test

- The documentation accurately describes the technical aspects of the product.
- The text, examples, and illustrations are technically correct.
- All technical information specified in the documentation plan is included.
- The procedures in all written and online documentation have been performed to ensure technical accuracy and completeness.

Marketing

- The documentation describes the product as it is specified in any marketing requirements document.
- The documentation is appropriate for the target audience.
- Proprietary information or information deemed inappropriate for publication outside the company is not included in the documentation.

Customer Support

- The documentation accurately describes the technical aspects of the product.
- The procedures in all written and online documentation have been performed to ensure technical accuracy and completeness.
- The documentation is appropriate for the target audience.
- The documentation includes information about how to obtain technical support.

Usability Testing

- The documentation is appropriate for the target audience.
- The documentation facilitates users' successful completion of tasks that the product supports.
- The instructional objectives of any tutorial material are met.

Publications Department

- The documentation conforms to current department and corporate standards and guidelines.
- The documentation is written in accordance with the documentation plan.
- The documentation is easy to read, well organized, and easy to use.
- The documentation is composed correctly: Tables and figures are clear and easy to read, no mistakes such as typographical errors, misspellings, and so on appear.

Legal

- The documentation properly uses all trademark and copyright conventions.
- All product names in the documentation are accurate and used correctly.

Review Meeting

The review meeting is chaired by the documentation manager or a designated representative. The sign-off reviewers designated in the documentation plan should attend the review meeting. Sign-off reviewers attending the review meeting are responsible for consolidating comments from their area into one copy and for resolving conflicting comments from different reviewers in their area before submission.

The writer or writing team leader should prepare an agenda of comment items. In an ideal situation where comments have been collected online before the meeting, you can distribute reports that list those comments that have been accepted without discussion, those comments that have been rejected, and those comments that are unresolved. Otherwise, mark on the review copies those comments that are unresolved or rejected.

The agenda for the meeting should include rejected comments that reviewers still want to have considered and unresolved comments.

The chairperson leads a discussion of each item on the agenda. Once the items have been resolved, the sign-off sheet should be distributed for representative signatures.

Often engineers and others feel that their comments regarding organization or writing style should receive as much weight as their comments on technical matters. Keep in mind, and make sure that they keep in mind, that *you* are the expert on publications just as *they* are the experts on coding or testing.

Internationalization and Globalization

If your company does a portion of its business internationally, you might need to localize or internationalize your documentation. *Internationalization* involves creating a "generic" document that can be easily translated into or used in many languages or cultures, converting any language-specific or culture-specific references into generic ones. *Localization* involves converting a document that is specific to a language or culture into one that is specific to a different language or culture. See Chapter 8, "Writing for an International Audience," for more specific details.

Some companies simply translate their documentation into specific languages. Other companies fully internationalize or localize the product and documentation. Although larger companies generally have a dedicated department to handle this process, others depend on the publications department.

Internationalization and localization is a large and complicated area that requires some expertise. Decisions about the process that affect publications include the following factors:

- How soon the localized versions need to be ready after the native-language product is finished. This schedule is often determined by marketing and sales concerns rather than publications resources.
- How changes to the documentation within the product release and in subsequent releases are tracked and passed on to the translators.
- How long the translation cycle takes, including review and revisions.
- How to handle exchanging documents between different formats, both hardware and software.

For smaller documents, you might want to have all languages in one document. For larger documents, you probably will want separate documents for each language. This decision must also take into account your company's manufacturing, product kitting, and inventory methods, and if your documentation is printed, the number of copies to be printed for each locale.

Online Documentation Considerations

Producing documentation for online presentation involves different concerns than the considerations for printed paper documentation. This section briefly discusses some issues you need to take into account, but it is by no means comprehensive. See Appendix E, "Recommended Reading," for books that deal with this subject in depth.

Many companies are turning to online presentation of their documentation to provide customers with easier access to their documentation, to save on printing and production costs, and to provide searchable linked information. If your company is considering such a move, research presentation and usability issues thoroughly.

See Chapter 5, "Online Writing Style," for detailed information about considerations for online presentation. The following sections contain summary information intended for publications departments new to online documentation.

Writing Issues

Because printed text has existed for hundreds of years, writers and readers instinctively know how to use it. Online text, however, is new and constantly changing. People writing and using online text do not have the combined knowledge that comes from generations of experimentation and example.

- **Writing style.** To design online information, writers often must change from the familiar writing style used for printed text to an unfamiliar, sometimes undefined, style used for online presentation. Some writers also often have a hard time going back and forth between the two writing styles.

- **Scheduling.** Time must be built into the schedule to enable writers and editors to develop their online writing style, and to transition between the printed and online formats.
- **Platform-specific versions.** Some information providers use a "write-once publish-twice" scheme that presents the same documentation online and in print. To most effectively take advantage of the benefits of online publishing, however, writers should write text specifically for that medium.

Content Issues

The following list presents some content issues related to online presentation.

- **Access.** You need to determine how the user will find and retrieve information, and what navigation aids you will provide.
- **Text format.** First, you must decide whether to duplicate information in both online and print formats, or whether you will provide some information only online and some only in print. If you choose the latter strategy, you need to decide the appropriate format for each type of information.

  One typical strategy is to provide task information or brief explanatory information online, and more in-depth information or background information in print.
- **Graphics.** If you decide to provide graphics in your online documentation, you must determine what format they will be in, how they will be included, and whether they will be linked.
- **Hyperlinks.** Some decisions to be made about hyperlinks include which information should be linked and whether links should be accessed only through text or also through graphics. You might also decide to limit the number of links you want in a given body of information. See Chapter 6, "Constructing Links," for more detailed information.
- **Page design.** Type size, page size, and page layout work differently online than in print. You must decide whether to optimize your design for online or print presentation or whether to use two separate designs.

Management Issues

The following list presents some management issues related to online presentation.

- **Scheduling.** Designing and planning online documentation is usually more time-consuming than print presentation because of the possibility of linked information and the ramifications of online display.
- **Authoring tools.** Online documentation usually requires a different set of authoring tools than your standard word processing or desktop publishing software. These tools sometimes involve separate products for authoring and for online viewing. Also, some web presentation tools do not provide the same range of formatting as authoring tools intended for print output, which can affect the way some information is presented.
- **Outside resources.** Support from other departments is usually required to a greater extent for online documentation than for print documentation. For example, you need to work more closely with graphic designers, product test, and the software integration team. When planning your online documentation strategy, make sure that the various groups have agreed to provide these resources.
- **Delivery mechanism.** Will your online documentation be part of the interface (as is usually the case with online help, for example) or the product code? Your deadlines and testing procedures will be affected by this decision.
- **Integration.** How will your documentation be connected to the product? You need to find out how the user will access the online information and how it will be installed and set up.
- **Testing.** Testing online documentation involves both the content of the documentation and the delivery media. Links that the user can click to go rapidly to cross-referenced information, for example, are one of the benefits of many online delivery products. Each link must be tested to make sure that it goes to the appropriate location. If the documentation is being delivered as part of the product, the online documentation must work correctly with the product code. Finally, the medium itself must be tested to make sure that it works on all supported platforms.
- **Legal.** Presenting information online usually means a change in how copyright and trademark issues are handled. Consult with your legal counsel.
- **Production.** If your online documentation will be produced on CD-ROM or DVD media, you must determine whether it will be included with the product software or on a separate disc. This decision can also affect your packaging.
- **ASCII text.** Often, last-minute product changes or additions are documented in a brief online file in ASCII text. Consider drafting some formats for this type of presentation.

Final Print Production

Activities associated with the final production of hard-copy documentation are as follows:

- Printing
- Binding
- Packaging

In larger companies, typically a separate production department handles these activities. The publications organization simply hands off the final camera-ready copy or electronic files. In smaller organizations, the publications department is often responsible for these activities.

This section provides only general information about printing and production processes. See Appendix E, "Recommended Reading," for sources of more thorough information on these subjects.

Deciding on a Strategy

Several factors influence the type of printing and packaging used for documentation. Some of these factors include the following:

- **Audience.** Is the documentation aimed at in-house engineers or commercial end users? This answer might determine your page layout or presentation. Under what conditions will your audience use the product? This answer might determine whether your books need to lie flat or whether you need to use a sturdier paper weight or binding method.
- **Competitors.** How are similar products printed or packaged?
- **Distribution method.** If you sell directly to your customers, packaging can be minimal, for example, an envelope or corrugated box. If you are selling to resellers, you probably want a more professional presentation.
- **Cost.** Your operations or marketing departments will probably include the documentation production cost limit when they set a profit level for the product.

Printing Methods

While documents can be reproduced in various ways, the two main printing methods are offset printing and photocopying. If you need fewer than 1000 copies of your manuals, you probably want to use photocopying. For 1000 to 4000 copies, regular sheet-fed offset is usually appropriate. For over 4500 copies, you will probably need to use a printer with a web press (one that uses rolls of paper rather than sheets).

Offset Printing

Offset printing provides higher quality than photocopying and might be your only choice if you are using color in illustrations or text or including photographs. Note that if you are producing camera-ready copy on a 300-dots per inch (dpi) laser printer, offset printing will not increase the quality of the output.

Offset printing requires a larger print quantity to be cost effective. Ordering a larger quantity just to take advantage of the lower price is usually not a good strategy, though, because you might be left with many copies of outdated manuals. Costs for offset printing are also influenced by other factors, including the use of color, the size of the page (influencing how much trimming is needed), and the type of paper and ink used.

Photocopying

Photocopying produces lower-quality output than offset printing, but it is more cost effective for smaller quantities. Other costs for photocopying might include special handling if you are using an odd size of paper or special paper. Many photocopying machines today can provide rudimentary binding as well as photocopying.

Binding Methods

Several types of binding are generally available. The most common binding types are as follows:

- Three-ring binders
- Wire-o
- Perfect
- Saddle-stitch

Consider the following factors when determining the binding method to use:

- Page size and number of pages
- Need for update insertions
- Frequency of users' access and the environment in which the book will be used
- Cost
- Customer preference

Three-Ring Binders

Three-ring binders hold standard 8.5-inch x 11-inch pages or other industry-standard page sizes. (For a price, you can also arrange for custom-size binders.) A big advantage to binders is that you can easily insert updates, change pages, and tabs. Covers can be slipped into plastic pockets in some binders on the front and spine, a feature that saves binder printing costs. However, three-ring binders take up a lot of room. Users often dislike binders because they are bulky and awkward to handle, and the rings can burst or rip pages.

Wire-O

Wire-o binding is much less expensive than three-ring binding, is smaller, and lies flatter. This binding offers much more variety in page size. However, updates are difficult to include and the wire prevents the book title from being printed on the spine. Wraparound covers are the workaround to this problem, but they are not very effective and add cost. Generally, users don't like wraparound covers and often can't figure out how to use them.

Perfect

Perfect binding is less expensive than either three-ring or wire-o binding, especially in large quantities. A perfect-bound book usually includes the book title on its spine, making the book easily identifiable on a shelf. The chief complaint about perfect-bound books in the past was that they did not lie flat, but lay-flat or flex bindings have remedied this problem. You cannot insert updates or change pages into perfect-bound books. Also, large page counts sometimes mean that pages might fall out over time.

Saddle-Stitch

Saddle-stitch binding is possible only for page counts up to 96 pages. While this binding is very inexpensive, you cannot insert updates or change pages into saddle-stitched books, and they are too thin to have a spine.

Packaging

The type of packaging you use should be determined in large part by how you sell your products and what your competition provides. If you sell products directly to your customers, you might not need as elaborate a packaging scheme as if you were to sell products through an external commercial vendor, where your package competes with others.

Direct-to-customer packaging could be as simple as a padded envelope or corrugated box.

The type of *commercial* packaging you choose could be influenced by several factors:

- The packaging your competitors produce and, therefore, the packaging your customers expect.
- The number of different pieces delivered. For example, do you have several disks, several manuals, a quick reference card, and a warranty card? Or do you have a single disk and a single manual?
- Your budget.

If you are new to dealing with commercial packaging, you might want to go through your printer to find a reputable packaging source. You can also go through a broker, who puts together a whole production package, finding and dealing with printing and packaging vendors, for a fee.

Once you find a packaging vendor, be sure to have the vendor produce prototypes of several packaging designs, and ask the pros and cons of each. You will probably have to provide the page counts for your manuals, and the number and type of other components in the package, before a realistic prototype can be produced.

Working With Outside Vendors

The best way to find a reputable printer or production broker is to ask people in your geographic area and industry. Ask printers or brokers about their experience with your type of product or about other jobs they have done for people in similar industries.

If your printing and packaging needs are relatively simple, you can probably use a printer who has less experience with your particular product area. However, you will be expected to provide the printer with camera-ready copy and all the information needed for the job.

If your needs are more complex or you have little experience in this area, you will probably need to spend more time finding a vendor who is right for you. Experienced printers can provide you with professional advice and samples. They can also explain the pros and cons of various printing methods, paper, ink, and so on, but you have to ask. Larger printers often have account representatives who can offer suggestions and get answers for you if the size of your account merits personal attention.

If you suspect your printing or packaging needs are very complex or you do not have either the expertise or the staff to investigate vendors, you might want to find a broker. Brokers quote a fee for your whole job and take on the responsibility of finding and dealing with vendors, procuring packaging samples, and so on.

Post-Production Considerations

After the documentation is delivered to production, your work is not done. You should have a plan in place to deal with last-minute product changes or inaccuracies in the documentation. Also, if the product documentation is likely to be revised, you should collect and maintain information that will help with the next version.

Handling Post-Production Revisions

Despite your best efforts, you might discover technical inaccuracies after the documentation is produced, either from omissions or from changes to the product. Therefore, allocations of monetary and staff resources to the project must continue even after the date the final documentation is delivered to production.

Typical ways to correct documentation inaccuracies include the following strategies:

- **Online files on the product disks.** These files can cover documentation inaccuracies and last-minute product revisions.
- **Online updates.** Depending on your online documentation delivery strategy, you might be able to update the documentation content between official product releases, for example, if your documentation is accessed through your company's web site. Make sure to consider how such changes are tracked and how customers can determine which version of the documentation they are viewing.
- **Replacement pages.** Errors in limited locations in printed manuals can be corrected through replacement pages that are inserted individually by users. These pages generally ship with the product and are accompanied by a card or insert listing all replacement pages.
- **Documentation update package.** A last-minute change to the product that has ramifications throughout the documentation might require too many change pages to make replacement pages feasible. In this case, you might want to include a documentation update package, which explains the exact changes that need to be made to specific pages, including the paragraph and line location and the information to be added or deleted. However, this solution should be a last resort.

Note – If your company sells directly to customers rather than through a third-party commercial distributor, you can issue change pages or update packages through your distribution channel or sales staff even after the product ships.

Make sure that customer support, marketing, localization, and sales personnel know about the inaccuracies and the steps taken to correct them.

Maintaining Project Continuity

Once a product has shipped, your first impulse might be to try to forget about it as soon as possible. However, your job will only be more difficult when you have to revise the documentation for the next version.

Some of the information you might want to save or document about a project includes:

- The location of online documentation files. Delete irrelevant files such as earlier versions of text or graphics. You might want to establish a central archive where final files are kept.
- While the details are still fresh in your mind, conduct a *post-mortem meeting* with the documentation team and write down a brief project review. Document anything that might be helpful, for example:
  - Processes that did not work as anticipated or could be modified to work better.
  - Controversies that arose. Even if they were resolved, having a record of each controversy could prove useful later.

- Issues that affected the schedule that might recur during the next cycle.
- Subject matter experts or other helpful project team members who were not official reviewers or project team members.

- Technical review comments from all official reviewers.
- Edits or review comments that you did not have time to incorporate for this version but that should appear in a future version.
- Product information that you did not have time to incorporate for this version but that should appear in a future version.

APPENDIX B

General Term Usage

This appendix lists and briefly explains how to correctly use terms that writers and editors frequently look up.

Note – Use the terms as stated in the table in this appendix, and capitalize and punctuate them as shown.

Certain terms might appear differently in products, applications, or other material you are documenting. For example, many companies say "log in," not "log on." However, if the audience you are writing for is accustomed to "log on," use that terminology. Another example is that of "root" and "superuser." Follow the style of the material you are documenting.

See also these sections in this guide for related guidelines and examples:

- "Capitalization" on page 25
- "Punctuation" on page 39 – Note especially the lists of prefixes in the "Hyphen" section.
- "Technical Abbreviations, Acronyms, and Units of Measurement" on page 35
- "Anthropomorphisms" on page 105 – Note especially the table of anthropomorphisms to avoid and suggested alternatives.
- "Idioms and Colloquialisms" on page 107 – Note especially the table of idioms and colloquialisms to avoid and suggested alternatives.
- "Phrasal Verbs" on page 110 – See this section for lists of acceptable phrasal verbs and phrasal verbs to avoid.
- Table 3–4 – See this table for a list of common redundancies to avoid and suggested alternatives.
- Chapter 14, "Documenting Graphical User Interfaces" – Note especially the sections on using GUI terminology, writing about windows, and writing about the web.

Note – Command names (`cd`, `man`), driver names (`ata`, `hme`), names of file formats (`gif`, `mpeg`), and so on generally do not appear in this table. See Appendix C, "Typographic Conventions," for related terms and examples and their typographic conventions.

TABLE B–1 Term Usage and Style

Term	Usage
Numbers/Symbols	
%	Use this symbol in tables and slides to conserve space. Use "percent" in text. Also indicates a user prompt.
#	This character is called a "comment mark" or the "superuser" prompt depending on the context. Do not use this character as an abbreviation for "number" or "pound." Use "no." or "lb" instead.
'	Do not use as an abbreviation for "foot." Use "ft" instead.
"	Do not use as an abbreviation for "inch." Use "in." instead.
0.25 inch (noun), 0.25-inch (modifier)	Hyphenate when used as a modifier.
0.25 inch, 0.5 inch	Do not use "1/4 inch" and "1/2 inch" except when referring to Allen wrenches and tape drives, for example, "1/2-inch tape drive."
1/4-inch Allen wrench	Put the measurement of the tool first. Note capitalization and unit of measure.
1U, 2U	Abbreviation for "1 rack unit," "2 rack units." The abbreviation "U" is singular, uppercase, with no space between the abbreviation and the numeral, for example, "2U." Abbreviation is *not* "RU."
2-post rack, 4-post rack	Use a numeral to refer to a post rack. Note hyphenation and lack of capitalization.
2-D (modifier)	Acronym for the standard use of "two-dimensional."
3-D (modifier)	Acronym for the standard use of "three-dimensional."
5.15.*x*	The last character represents a variable.
10BASE2, 10BASE5	Note capitalization and lack of hyphenation.
10BASE-T, 10BASE-TX	Note capitalization and hyphenation.
100BASE-TX, 100BASE-FX, 100BASE-T4	Note capitalization and hyphenation.

TABLE B–1 Term Usage and Style *(Continued)*

Term	Usage
A	
above	Do not use to refer to the location of another piece of information. Instead, use "previous" or "preceding," or refer to the specific section heading or figure number, for example.
AC/DC converter	Note capitalization and the slash. Do not spell out "AC/DC."
adapter	Note spelling.
add-on	noun, modifier
address bus	Refers to a bus type, not a bus name.
affect	Verb meaning "to change or influence something." For example, "The style setting affects the appearance of the paragraph." See also "effect."
after	Do not use to refer to the location of another piece of information. Instead, use "next" or "following," or refer to the specific section heading or figure number, for example.
airflow	noun
align to	Do not use. Use "align with."
align with	Use instead of "align to."
Allen wrench	Note capitalization.
allow	Use only when discussing permission. For example, "Write access allows the user to modify the file." When discussing capabilities, use "enables" or rewrite the sentence. For example, "The Edit menu options enable you to modify the document" or "Use the Edit menu options to modify the document." Also applies to "lets" and "permits."
alphanumeric	modifier
a.m.	Abbreviation for "ante meridiem" (morning). Note periods in "a.m."
and/or	Do not use. If you mean "or," write "or." If you mean "and," write "and." If you mean that any or all of the things that are named might be affected, say so. For example, "Using the Edit menu, you can cut, paste, or cut and paste text or graphics."
antialiasing (noun), antialiased (modifier)	Note lack of hyphenation.
antiglare	modifier
antistatic	modifier
antistatic strap	Use instead of "grounding strap" or "ESD wrist strap."

TABLE B–1 Term Usage and Style (*Continued*)

Term	Usage
append (verb)	Use instead of "postpend." Use to denote the action of placing data at the end.
ASCII	Note capitalization. Do not spell out.
assembly language (noun), assembly-language (modifier)	Hyphenate when used as a modifier.
ATA-2, ATA-3...	Note hyphenation.
audio-in	noun, modifier
audio-out	noun, modifier
autoboot	verb, modifier
autoconfigure	verb
autodetect	verb, modifier
autoloader	noun
automount	verb
autotermination	noun, modifier
B	
b	Abbreviation for "bit."
B	Abbreviation for "byte."
back end (noun), back-end (modifier)	Hyphenate when used as a modifier.
back panel (noun, modifier)	Use instead of "rear panel."
backplane	noun
backplate	noun, modifier
backquote	noun, modifier
backslash	noun, modifier
backspace	verb, modifier
Backspace key	Note capitalization.
backup (noun, modifier), back up (verb)	Note lack of hyphenation.
backward	Note lack of final "s."

TABLE B–1 Term Usage and Style *(Continued)*

Term	Usage
bandwidth	noun, modifier
base plate	noun
baud rate	Often incorrectly assumed to indicate the number of bits per second (bps) transmitted, baud rate actually measures the number of events, or signal changes, that occur in one second. In most instances when "baud rate" is used, the correct term is "bps." For example, a so-called 9600-baud modem that encodes 4 bits per event actually operates at 2400 baud, but it transmits 9600 bits per second (2400x4 bits per event) and thus is correctly called a 9600-bps modem. Check your source material before using the term "baud rate."
before	Do not use to refer to the location of another piece of information. Instead, use "previous" or "preceding," or refer to the specific section heading or figure number, for example.
below	Do not use to refer to the location of another piece of information. Instead, use "next" or "following," or refer to the specific section heading or figure number, for example.
benchmark	noun, modifier
bidirectional	modifier
big-endian (noun, modifier)	Note hyphenation and lack of capitalization.
BIST	Acronym for "built-in self-test."
bisynchronous	modifier
bit block	noun, modifier
bit field (noun), bit-field (modifier)	Hyphenate when used as a modifier.
bitmap	noun, modifier
bitstring	noun, modifier
boot (verb)	Use instead of "boot up" or "bootup." Use only to indicate a software boot. Use "power on" for hardware.
BootBus (noun)	Note one word. Note capitalization, which is an exception to bus naming conventions.
boot PROM (noun)	Note capitalization. Do not spell out "PROM."
bootstrap (verb)	Refers to loading the system software into memory and starting it.
boot up, boot-up	Do not use. See "boot."
Bourne shell	Note capitalization.

TABLE B–1 Term Usage and Style *(Continued)*

Term	Usage
bpi	Abbreviation for "bits per inch."
Bpi	Abbreviation for "bytes per inch."
bps	Abbreviation for "bits per second."
Bps	Abbreviation for "bytes per second."
breakpoint	noun, modifier
bring the system down	Do not use. Write "cause the system to fail," "shut down the system," or "power off the system," depending on the meaning.
bring up	Do not use. Write "power up the system," "start the system," "turn on the machine," or "turn on the power to the system," or other text, depending on the meaning.
broadband	modifier
browse	Use instead of "explore."
browser	Use instead of "web browser" or "web browser window." Use "browser" if you are describing an action, such as "Start the browser."
browser interface	Use instead of "browser user interface" or "BUI." Use "browser interface" to distinguish between two interfaces. For example, "Access the configuration and monitoring software from a browser interface or a command-line interface."
browser user interface (BUI)	Do not use. Use "browser interface."
BUI (browser user interface)	Do not use. Use "browser interface."
built-in	modifier
built-in self-test (BIST)	Note hyphenation and lack of capitalization of spelled-out form.
bundled	Do not use in external documents. Describe bundled products (for example, the CDE interface) as part of the Solaris OS. See also "co-packaged."
buses	Note spelling.
bus nexus	modifier
byte stream (noun), byte-stream (modifier)	Hyphenate when used as a modifier.

TABLE B–1 Term Usage and Style *(Continued)*

Term	Usage
C	
c	Abbreviation for “centi” (prefix).
°C	Abbreviation for “degrees Celsius.” Note no space between the abbreviation and the numeral, for example, 4°C.
callback	noun, modifier
can	Use to indicate the power or the ability to do something. For example, “See if you can log in to the system.” See also “may” and “might.”
canceling	Note no double “l.”
cannot	Note spelling.
card cage	noun
case-sensitive	modifier
CD	Acronym for “compact disc.” Note spelling of “disc.”
CD-digital audio format	Note hyphenation and capitalization.
CD-ROM (noun, modifier)	Acronym for “compact disc read-only memory.” Note spelling of “disc.” Use to refer generically to CD-ROM media, for example, “The software is distributed on CD-ROM discs.” If referring to a specific compact disc for installation or other purposes, “CD” is acceptable, for example, “Load the CD into the CD-ROM drive.” If referring to CD-ROM media that contain wholly or mostly music, add the adjective “audio,” as in “audio CD-ROM.”
Celsius	Use instead of “centigrade.”
centerplane (noun)	Note lack of hyphenation.
centigrade	Do not use. Use “Celsius.”
check box (noun)	Note that some interface standards might spell this term as one word.
check-in (noun, modifier), check in (verb)	Hyphenate when used as a noun or modifier.
check-out (noun, modifier), check out (verb)	Hyphenate when used as a noun or modifier.
chipset	noun
CIFS	Abbreviation for “common Internet file system.”
classpath	Note lack of hyphenation.

TABLE B–1 Term Usage and Style *(Continued)*

Term	Usage
CLI	Abbreviation for "command-line interface."
client	Use only when talking about the relationship with a server. Do not use to refer to a person.
client-server	Use when describing a relationship between a client and a server. For example, "This network is based on the client-server model."
coaxial	modifier
codec	Acronym for "coder/decoder" and "compression/decompression." Define as appropriate to context.
code name	noun
codeset	noun, modifier
color map	noun, modifier
.com, dot-com	Use `.com` when referring to the suffix itself. Also use .com in appropriate trademarked terms, or when your document requires no variations on the term ".com." Use dot-com when referring to companies ("dot-coms"), when using as a verb ("to dot-com," "dot-commed," "dot-comming"), or when the term is the first word in a sentence or heading.
COM1, COM2…	Acronym for "communications port." You can also use the terms "serial port A, serial port B" and so on.
combo box	Do not use in user documentation. Name the box instead.
command line (noun), command-line (modifier)	Hyphenate when used as a modifier.
command-line interface (CLI)	Note hyphenation and lack of capitalization of spelled-out form.
common Internet file system (CIFS)	Note capitalization.
compact disc (CD)	Note lack of capitalization of spelled-out form. Note spelling of "disc."
compact disc read-only memory (CD-ROM)	See "CD-ROM."
CompactPCI (cPCI)	Note capitalization of the spelled-out noun (cPCI is a modifier).
compared with	Use instead of "vs." or "versus."
comprise	Avoid using "comprise." Use "contain" or "include" instead. Do not use "comprised of" when you mean "composed of."
configurable	Note spelling.

TABLE B–1 Term Usage and Style *(Continued)*

Term	Usage
connect (verb)	Use "plug" only in the context of "plug and play."
connector (noun)	Use instead of "plug."
connects to	Use instead of "connects with."
context-sensitive	modifier
controller	Note spelling.
Coordinated Universal Time (UTC)	Note capitalization. Note spelling of "UTC."
coprocessor	noun, modifier
coroutine	noun, modifier
cPCI (modifier)	Abbreviation for "Compact PCI" (noun). Note capitalization.
CPU	Do not spell out.
cross-assembler	noun, modifier
cross-compiler	noun, modifier
cross-reference	noun, modifier, verb
cross-section (noun), cross-sectional (modifier)	Note hyphenation. Note change for modifier.
currently	Use only in a document that you know will be updated regularly, for example, in release notes. You can also use a specific date, for example, "August 2009."
C shell	Note capitalization.
D	
d	Abbreviation for "deci" (prefix).
da	Abbreviation for "deka" (prefix).
D/A (modifier)	Abbreviation for "digital-to-analog." Note the slash in "D/A."
daemon	Note spelling. Pronounced "demon."
DAS	Abbreviation for "direct attached storage."
data	Although in the pure Latin form this noun is plural, the most common industry usage is in the singular. For example, "The data is available."
database	noun, modifier
data center (noun)	Note two words.

TABLE B–1 Term Usage and Style *(Continued)*

Term	Usage
dataless	Define this word on first reference, and include it in a glossary if your book has one.
data path (noun)	Note two words.
data set (noun)	Note two words.
datasheet	noun, modifier
data store	Note two words.
data stream (noun, modifier)	Note two words.
data type (noun)	Note two words.
daughterboard	noun, modifier
dB	Abbreviation for "decibel."
D-connector	noun
deconfigure	Do not use. Use "unconfigure."
default	Define this word narrowly and use it consistently.
deinstall	Do not use. Use "uninstall."
depress	Do not use when providing a keyboard instruction. Use "press."
depth cueing	noun
deselect	verb
design	Do not use. Do not write, for example, "XYZ is designed to search for files." If XYZ is designed to search, assume that it does. Write "XYZ searches for files."
desire	Do not use. Use "want."
desktop	Use only when referring to a specific piece of hardware. Otherwise, use "system" or "host."
device	Define this word narrowly and use it consistently.
device driver (noun, modifier)	Note lack of hyphenation.
dialog	Do not use. Use "dialog box."
dialog box	Use instead of "dialog."
dial up (verb), dial-up (modifier)	Hyphenate when used as a modifier.
die	Do not use. Use "fail."

TABLE B–1 Term Usage and Style (Continued)

Term	Usage
digital-to-analog (D/A) (modifier)	Note hyphenation and lack of capitalization. Note the slash in "D/A."
digital video disc (DVD)	Note lack of capitalization of spelled-out form and spelling of "disc."
DIMM	Acronym for "dual inline memory module." Do not spell out.
direct attached storage (DAS)	Note lack of capitalization of spelled-out form. Use "attached," not "attach."
directory name (noun), *directory-name* (variable)	Note hyphenation when used as a variable.
disappear	Do not use. A window does not "disappear." Use "dismiss" instead.
disc	Note spelling for a compact disc, a laser disc, an optical disc, or a video disc.
disk	Use to refer to a hard disk.
disk drive	Use instead of "hard disk drive" or "fixed disk." Depending on the product, "hard drive" is also acceptable.
diskette	Use instead of "floppy" or "floppy disk."
diskette drive	Use instead of "floppy drive" or "floppy disk drive."
diskfull	Define this word on first reference and include it in a glossary if your book has one.
diskless	Define this word on first reference and include it in a glossary if your book has one.
DNS	Abbreviation for "domain name server," "domain name service," and "Domain Name System." Define as appropriate to context.
domain name server (DNS)	Note lack of capitalization of spelled-out form.
domain name service (DNS)	Note lack of capitalization of spelled-out form.
Domain Name System (DNS)	Note capitalization of spelled-out form.
dot-com, .com	Use dot-com when referring to companies ("dot-coms"), when using as a verb ("to dot-com," "dot-commed," "dot-comming"), or when the term is the first word in a sentence or heading. Use `.com` when referring to the suffix itself. Also use .com in appropriate trademarked terms, or when your document requires no variations on the term ".com."
double click (noun), double-click (verb, modifier)	Hyphenate when used as a verb or modifier.

TABLE B–1 Term Usage and Style *(Continued)*

Term	Usage
double-density	modifier
double-sided	modifier
doubleword (noun), double-word (modifier)	Hyphenate when used as a modifier.
download	verb, modifier
downtime	noun
DR (noun, modifier)	Abbreviation for "dynamic reconfiguration."
drag and drop (verb), drag-and-drop (modifier)	Hyphenate when used as a modifier.
drive 0, drive 1...	Use a numeral to refer to a drive number.
drive A:, drive B:	Use a capital letter followed by a colon to refer to a drive letter.
dual-access	modifier
dual-density	modifier
dual inline memory module (DIMM)	Do not use the spelled-out form.
dump file	noun
DVD	Acronym for "digital video disc." Note spelling of "disc."
dynamic reconfiguration (DR) (noun, modifier)	Note lack of capitalization of spelled-out form when used in a generic sense.
E	
EBus	Note capitalization.
Ecache	Note capitalization.
ECC	The most common spelled-out forms are "error correcting code" and "error checking and correction." Define as appropriate to context.
e-commerce	Note hyphenation and lack of capitalization.
edge-triggered interrupt	noun
effect (noun)	Means a result or consequence. For example, "The style setting has an effect on the appearance of the paragraph." See also "affect."
e.g.	Do not use. Use "for example."
EIDE connector	Note capitalization.

TABLE B–1 Term Usage and Style *(Continued)*

Term	Usage
ejection lever	Use instead of "ejector lever."
ejector lever	Do not use. Use "ejection lever."
electric shock	Use instead of "electrical shock."
electrical shock	Do not use. Use "electric shock."
electrostatic discharge (ESD)	Note lack of capitalization of spelled-out form.
email	noun, modifier (not a verb)
enables	See "allow."
-enabled	Avoid. Use text such as "works with" or "is compatible with" instead.
endianness	noun
endpoint	noun, modifier
end user (noun), end-user (modifier)	Hyphenate when used as a modifier. Use only if you need to distinguish between a developer (your user) and the person (the end user) who will be using the product the developer designs.
ensure that	Include the word "that" when introducing a restrictive clause.
enter	To avoid confusion with the Enter key, do not use. Use "type" if the action involves the actual typing of data.
environment	When using this word outside a product name context, define this word narrowly and use it consistently.
EOL	modifier (not a verb)
Eport	Acronym for "expansion port." Note capitalization of "Eport."
error checking and correction (ECC)	Define as appropriate to context. Note lack of capitalization and hyphenation of spelled-out form.
error correcting code (ECC)	Define as appropriate to context. Note lack of capitalization and hyphenation of spelled-out form.
ESD	Abbreviation for "electrostatic discharge."
ESD wrist strap	Do not use. Use "antistatic strap."
etc.	Do not use. Use more explicit text, which also solves problems with sentence-ending punctuation. For example, do not write "Mail Tool enables you to compose email messages, respond to email messages, etc." Instead, write "Mail Tool enables you to compose email messages, respond to email messages, and perform other mail administration tasks."

TABLE B–1 Term Usage and Style *(Continued)*

Term	Usage
Ethernet	Note capitalization.
example (noun, modifier)	Use instead of "sample" in reference to programming code.
expansion bus	noun
expansion memory bus interface	noun
expansion port (Eport)	Note lack of capitalization of spelled-out form. Note capitalization of "Eport."
explore	Do not use when you mean "navigate" the web.
externally initiated reset (XIR)	Note lack of capitalization of spelled-out form. Note spelling of "XIR."
F	
F	Abbreviation for "farad."
°F	Abbreviation for "degrees Fahrenheit." Note no space between the abbreviation and the numeral, for example, "32°F."
faceplate	noun
fail	Use instead of "die."
failback (modifier)	Note lack of hyphenation.
failover (noun), fail over (verb)	Note lack of hyphenation.
fast-on	modifier
Fast Ethernet (noun, modifier)	Note capitalization.
Fast SCSI (noun, modifier)	Note capitalization.
Fast Wide SCSI (noun, modifier)	Note capitalization and no slash.
Fast Wide UltraSCSI (noun, modifier)	Note capitalization and no slash.
FC-AL	Abbreviation for "Fibre Channel-Arbitrated Loop." Note hyphenation.
FCC Class A, FCC Class B (modifier)	Note capitalization.
FCode	Note capitalization.
feedback	noun, modifier
female connectors	Do not use. Name the connectors instead.
fiber optics (noun), fiber-optic (modifier)	Hyphenate when used as a modifier. Note lack of final "s" for modifier.

TABLE B–1 Term Usage and Style *(Continued)*

Term	Usage
Fibre Channel	Note spelling.
Fibre Channel-Arbitrated Loop (FC-AL)	Note capitalization and hyphenation.
field-replaceable unit (FRU)	Note spelling of "replaceable." Acronym pronounced "fru" so the preceding indefinite article should be "a," not "an."
file name (noun, modifier), *filename* (variable)	Note lack of hyphenation.
file server (noun, modifier)	Note lack of hyphenation.
file sharing (noun), file-sharing (modifier)	Hyphenate when used as a modifier.
file system (noun, modifier), *filesystem* (variable)	Note lack of hyphenation.
filler panel	Use instead of "filler plate."
firewall (noun)	Note one word.
firmware	noun
fixed disk	Do not use. Depending on the product, use "disk drive" or "hard drive."
fixed length (noun), fixed-length (modifier)	Hyphenate when used as a modifier.
fixed point (noun), fixed-point (modifier)	Hyphenate when used as a modifier.
flash PROM	Note capitalization. Do not spell out "PROM."
flat-blade	modifier
floating point (noun), floating-point (modifier)	Hyphenate when used as a modifier.
floppy	Do not use. Use "diskette."
floppy disk	Do not use. Use "diskette."
floppy disk drive	Do not use. Use "diskette drive."
floppy drive	Do not use. Use "diskette drive."
flowchart	noun
following	Use instead of "below" or "after" to refer to the location of another piece of information. Consider using "next" or referring to the specific section heading or figure number, for example.

TABLE B–1 Term Usage and Style *(Continued)*

Term	Usage
follow-up (noun, modifier), follow up (verb)	Hyphenate when used as a noun or modifier.
for example	Use instead of "e.g."
Fortran	Note the capitalization of this acronym for "formula translation."
frame buffer	noun
framework	noun
front end (noun), front-end (modifier)	Hyphenate when used as a modifier.
front panel	noun
FRU	Acronym for "field-replaceable unit." Pronounced "fru" so the preceding indefinite article should be "a," not "an."
ft	Abbreviation for "foot." Do not use the symbol (′).
ft lb	Abbreviation for "foot-pound."
full-duplex	modifier
G	
g	Abbreviation for "gram."
G	Abbreviation for "giga" (prefix). In computer terminology, "G" represents 2^{30}, or 1,073,741,824.
Gb	Abbreviation for "gigabit."
GBIC	Abbreviation for "gigabit interface converter." Note spelling of the abbreviation of this generic term.
Gbit/sec, Gbps in tables only	Acronym, abbreviation for "gigabits per second." In tables, footnote the first occurrence of "Gbps" and state that it stands for "gigabits per second."
Gbyte; G, GB in tables only	Acronym, abbreviations for "gigabyte."
Gbyte/sec, GBps in tables only	Acronym, abbreviation for "gigabytes per second." In tables, footnote the first occurrence of "GBps" and state that it stands for "gigabytes per second."
Gcache (noun)	Note capitalization.
GENERIC configuration kernel	Note capitalization.

TABLE B–1 Term Usage and Style *(Continued)*

Term	Usage
geo-	Use a hyphen with this prefix for "geo-specific" and "geo-neutral," but *not* for "geocentric."
geographic region	Use instead of "geographical region."
GHz	Abbreviation for "gigahertz."
GigabitEthernet (noun)	Note one word, capitalization, and lack of hyphenation.
gigabit interface converter (GBIC)	Note capitalization. Note spelling of the abbreviation for this generic term.
GNOME	Pronounced "guh-nome" so the preceding indefinite article should be "a," not "an."
graphics card	Do not use. Use "video display device."
grayed-out	modifier
gray scale (noun), gray-scale (modifier)	Hyphenate when used as a modifier.
grounding strap	Do not use. Use "antistatic strap."
GX frame buffer	Note capitalization.
H	
HA (noun)	Abbreviation for "high availability."
half-duplex	modifier
half-height	modifier
halftone	noun, modifier
halfword (noun), half-word (modifier)	Hyphenate when used as a modifier.
handheld	modifier
hands-on	modifier
hard-coded	modifier
hard copy (noun), hard-copy (modifier)	Hyphenate when used as a modifier.
hard disk	Do not use. Use "disk."
hard disk drive	Do not use. Depending on the product, instead use "hard drive" or "disk drive."

TABLE B–1 Term Usage and Style *(Continued)*

Term	Usage
hard drive	Depending on the product, use "hard drive" or "disk drive."
hardwire (verb), hardwired (modifier)	Note "d" at end of modifier.
heat sink	noun
help, Help	Do not use to refer to "online help."
hertz (Hz)	Note lack of capitalization of spelled-out form.
hex (modifier)	Acronym for "hexadecimal."
high availability (HA) (noun)	Note two words. Note lack of capitalization and hyphenation of spelled-out form of the generic feature.
high-availability	modifier
high density (noun), high-density (modifier)	Hyphenate when used as a modifier.
high end (noun), high-end (modifier)	Hyphenate when used as a modifier.
high level (noun), high-level (modifier)	Hyphenate when used as a modifier.
high order (noun), high-order (modifier)	Hyphenate when used as a modifier.
high performance (noun), high-performance (modifier)	Hyphenate when used as a modifier.
high priority (noun), high-priority (modifier)	Hyphenate when used as a modifier.
high resolution (noun), high-resolution (modifier)	Hyphenate when used as a modifier.
high speed (noun), high-speed (modifier)	Hyphenate when used as a modifier.
hit	Do not instruct a person to "hit" anything, including computer keys. Use "press."
hits (noun)	Use "hits" to refer to the number of times a particular piece of information appears when searching the web. See also "search results."
home page	noun
host (noun, verb)	When used as a noun, "host" typically means a computer connected to a network.

TABLE B–1 Term Usage and Style (Continued)

Term	Usage
host ID (noun), *hostID* (variable)	Note capitalization and lack of hyphenation.
host name (noun), host-name (modifier), *hostname* (variable)	Hyphenate when used as a modifier.
hot key (noun, verb), hot-key (modifier)	Hyphenate when used as a modifier.
hotline	noun, modifier
hot-plug, hot-plugging (verb); hot-pluggable, hot-plugged (modifier)	Note hyphenation.
hot-spare (noun, modifier)	Note hyphenation.
hot-swap, hot-swapping (verb); hot-swappable, hot-swapped (modifier)	Note hyphenation.
hypertext, hypertext link	Avoid. Use "link" instead.
Hz	Abbreviation for "hertz."
I	
I^2C	Abbreviation for "inter-integrated circuit."
i.e.	Do not use. Use "that is."
I18N	Limit this abbreviation for "internationalization" to internal documents.
ID, IDs (noun)	Note capitalization of single and plural forms of this acronym for "identifier."
if	This term can sometimes be replaced with "whether." If you can substitute "whether or not" for the word "if," use "whether" rather than "if." See also "when."
in.	Abbreviation for "inch." The abbreviation ends with a period. Do not use the symbol (") to mean "inch."
indexes	Use instead of "indices."
indices	Do not use. Use "indexes."
information on	Avoid. Use "information about."
inline	adverb, modifier
inode	Note lack of hyphenation. Pronounced "eye-node."
in order to	Avoid. Use "to."

TABLE B–1 Term Usage and Style *(Continued)*

Term	Usage
input	noun, modifier (not a verb)
input/output	See "I/O."
install	verb
installation	noun, modifier
instructions on	Use instead of "instructions about."
instruction set architecture (ISA)	Note lack of capitalization of spelled-out form.
interdomain	Note lack of hyphenation.
inter-integrated circuit (I^2C)	Note hyphenation and lack of capitalization of spelled-out form.
Internet	Note capitalization.
interoperability	Define narrowly in a network or Internet context. Otherwise, use "works with" or "is compatible with."
interpacket gap (IPG)	Note lack of capitalization and hyphenation of spelled-out form.
intranet	Note the lack of capitalization for an internal TCP/IP network.
invoke	Use only if no other word accurately describes the action. Usually, you can substitute "run," "start," or "call."
I/O (noun, modifier)	Abbreviation for "input/output." Note the slash. Is singular, for example, "I/O is," not "I/Os are."
IPG	Abbreviation for "interpacket gap."
IPMP	Abbreviation for "IP Network Multipathing." Note spelling of abbreviation. Note capitalization of spelled-out form.
IP Network Multipathing (IPMP)	Note capitalization. Note spelling of "IPMP."
ips	Abbreviation for "inches per second."
IPS	Abbreviation for "instructions per second" and "interrupts per second." Define as appropriate to context.
ISA	Acronym for "instruction set architecture."
ISO *xxxx* specifications (noun, modifier)	Note lack of hyphenation.

TABLE B–1 Term Usage and Style *(Continued)*

Term	Usage
its, it's	Without an apostrophe, "its" is the possessive form of the pronoun "it." For example, "The site describes the eMetrics program and its accompanying utilities." With an apostrophe, "it's" is an abbreviation for "it is." Do not use this contraction because of the potential confusion with "its." Instead, use "it is." For example, "This package is included because it is required for backward compatibility."
J	
Joint Test Action Group (JTAG) (noun, modifier)	Note capitalization.
K	
k	Abbreviation for "kilo" (prefix).
K	Abbreviation for "kelvin" (unit of absolute temperature). In computer terminology, "K" represents 2^{10}, or 1024.
K, KB	Abbreviations for "kilobyte." Use in tables only. Otherwise, use "Kbyte" or "kilobyte."
Kb	Abbreviation for "kilobit." Use in tables only. Otherwise, use the spelled-out form."
Kbit/sec, Kbps in tables only	Acronym, abbreviation for "kilobits per second." In tables, insert a footnote at the first occurrence of "Kbps" and state that it stands for "kilobits per second."
Kbyte	Acronym for "kilobyte."
Kbyte/sec, KBps in tables only	Acronym, abbreviation for "kilobytes per second." In tables, insert a footnote at the first occurrence of "KBps" and state that it stands for "kilobytes per second."
Kerberos security mechanism	Note capitalization.
keyboard	noun, modifier
keyboard/mouse port (noun)	Note the slash.
keylock	noun
keymap	noun, modifier
keypad	noun, modifier
keystore	noun

TABLE B–1 Term Usage and Style (Continued)

Term	Usage
keystroke	noun, modifier
keyswitch	noun
keyword	noun, modifier
kg	Abbreviation for "kilogram."
kHz	Abbreviation for "kilohertz."
km	Abbreviation for "kilometer."
Korn shell	Note capitalization.
kV	Abbreviation for "kilovolt."
kW	Abbreviation for "kilowatt."
kWh	Abbreviation for "kilowatt-hour."
L	
L1-A	Note capitalization.
L10N	Limit this abbreviation for "localization" to internal documents.
labeled	Note no double "l."
label-side up	modifier
laser disc	Note spelling.
launch	Do not use. Use "start."
lb	Abbreviation for "pound." Do not use the symbol # to mean "pound."
least significant	noun, modifier
LED colors	Note capitalization. Standard LED colors are "amber," "green," and "red." Do not use "yellow."
left side	Use instead of "left-hand side."
left-justified	modifier
let	See "allow."
Level 2 cache	Note capitalization.
level-triggered interrupt	noun
line-in, line-out	Both noun and modifier
link (noun, verb)	When used as a noun, use instead of "hypertext" or "hypertext link."

TABLE B–1 Term Usage and Style *(Continued)*

Term	Usage
link-fail state	noun
link integrity test	noun
link-pass state	noun
link test disabled	modifier
link test enabled	modifier
link test function	noun
link test pulses	noun
Linux (noun, modifier)	Note capitalization of this operating system.
list file	noun
little-endian (noun, modifier)	Note hyphenation and lack of capitalization.
lockup (noun, modifier), lock up (verb)	Note lack of hyphenation.
log file	noun
login (noun, modifier)	Note lack of hyphenation.
log in to (verb)	Use instead of "log into."
log into	Do not use. Use "log in to."
logoff, log off	Do not use unless these terms appear in an application that you are documenting. Use "logout" (noun, modifier) and "log out of" (verb).
logon, log on	Do not use unless these terms appear in an application that you are documenting. Use "login" (noun, modifier) and "log in to" (verb).
logout (noun, modifier), log out of (verb)	Note lack of hyphenation.
look and feel	noun
loopback	noun, modifier
lookup (noun), look-up (modifier), look up (verb)	Hyphenate when used as a modifier.
loosen or tighten captive screws	Use instead of "unscrew" or "screw."
low	See related terms under "high" for correct usage.
lowercase	noun, modifier, verb
lower left (modifier)	Use instead of "lower left-hand."

TABLE B–1 Term Usage and Style *(Continued)*

Term	Usage
lower right (modifier)	Use instead of "lower right-hand."
M	
µ	If your authoring environment does not support this symbol for "micro," spell out the measurement.
µA	Abbreviation for "microampere."
µF	Abbreviation for "microfarad."
µg	Abbreviation for "microgram."
µm	Abbreviation for "micrometer."
µs	Abbreviation for "microsecond."
µV	Abbreviation for "microvolt."
µW	Abbreviation for "microwatt."
m	Abbreviation for "meter" and "milli" (prefix).
M	Abbreviation for "mega" (prefix). In computer terminology, "M" represents 2^{20}, or 1,048,576.
mA	Abbreviation for "milliampere."
machine	Use only if a more specific hardware type is not appropriate.
machine-readable	modifier
mainframe	noun, modifier
main menu	Note lack of capitalization when used generically.
male connectors	Do not use. Name the connectors instead.
man page	The "man" part of this term when used generically is treated as a standard English word and is capitalized in headings and at the beginning of a sentence. For example, "Man pages are very helpful." However, the `man` command appears in monospace font and is lowercase.
mass storage (noun), mass-storage (modifier)	Hyphenate when used as a modifier.
master I/O controller	Note capitalization and the slash.
matrices	Do not use. Use "matrixes."
matrixes	Plural of "matrix." Use instead of "matrices."

TABLE B–1 Term Usage and Style (*Continued*)

Term	Usage
may	Use only when granting permission. For example, "You may use either uppercase or lowercase letters." Use "can" to indicate the power or ability to do something. Use "might" to indicate a possibility.
Mb	Abbreviation for "megabit." Use in tables only. Otherwise, use the spelled-out form.
MB	Abbreviation for "megabyte." Use in tables only. Otherwise, use "Mbyte" or "megabyte."
Mbit/sec, Mbps in tables only	Acronym, abbreviation for "megabits per second." In tables, footnote the first occurrence of "Mbps" and state that it stands for "megabits per second."
MBus	Note capitalization.
Mbyte	Acronym for "megabyte."
Mbyte/sec, MBps in tables only	Acronym, abbreviation for "megabytes per second." In tables, footnote the first occurrence of "MBps" and state that it stands for "megabytes per second."
MDB	The most common spelled-out forms are "multiple device boot" and "Modular Debugger." Define as appropriate to context.
memory controller	Note lack of capitalization.
menu-driven	modifier
metadata	noun, modifier
metadevice	noun, modifier
metadisk	noun, modifier
metafile	noun, modifier
Meta key	Note capitalization.
mF	Abbreviation for "millifarad."
MHz	Abbreviation for "megahertz."
mice	Do not use. Use "mouse devices" to refer to more than one mouse.
microcode	noun, modifier
micro-D	modifier
microprocessor	noun
midrange	noun, modifier

TABLE B–1 Term Usage and Style (Continued)

Term	Usage
might	Use to indicate a possibility. For example, "You might need to use another mouse." See also "can" and "may."
might want to	Do not try to read minds. Write "If you want to exit from the application, click Exit."
Modular Debugger (MDB)	Note capitalization. Note spelling of "MDB."
monochrome	modifier
monospace	modifier
-most	Do not use with directional words such as "left" or "top." Use phrases such as "on the left" or "at the far left" instead.
most significant	noun, modifier
motherboard	noun
mount	Define this word narrowly and use it consistently.
mount point (noun), mount-point (modifier)	Hyphenate when used as a modifier.
mouse devices	Use instead of "mice."
multi-master	Note hyphenation.
multi-owner	Note hyphenation.
MultiPack (noun)	Note capitalization and lack of hyphenation.
multiple address spaces	noun
multiple device boot (MDB)	Do not use initial capitalization for the spelled-out form unless the term appears that way on a label or screen.
multiplexer (MUX)	Not "multiplex." Note spelling of both the term and acronym.
mV	Abbreviation for "millivolt."
mW	Abbreviation for "milliwatt."
MW	Abbreviation for "megawatt."

TABLE B–1 Term Usage and Style *(Continued)*

Term	Usage
N	
name service	Refers specifically to DNS (domain name service).
namespace	noun, modifier
naming service	Generic term used with DNS, NIS, NIS+, or LDAP, for example, "the DNS, NIS, NIS+, and LDAP naming services are ..." or "the DNS and NIS naming services are...."
navigate	Use instead of "explore" or "surf."
NEBS	Acronym for "Network Equipment Building Systems."
needle-nose pliers	Not "needle-nosed pliers."
NetBIOS	Note capitalization.
Network Equipment Building Systems (NEBS)	Note capitalization.
network file system (NFS)	Note lack of capitalization of spelled-out form.
newline	noun, modifier
next	Use instead of "before" or "after" to refer to the location of another piece of information. Consider using "following" or referring to the specific section heading or figure number, for example.
NFS	Abbreviation for "network file system."
NIS+	This abbreviation for Network Information Service Plus is pronounced "en-eye-ess plus," so precede it with "an," not "a."
no. (noun, modifier)	Abbreviation for "number." Note the period. Do not use the symbol # to mean "number."
No. 2 Phillips screwdriver	Note capitalization and the period. Do not use the "-hand" suffix after "Phillips."
node name (noun), *nodename* (variable)	Note lack of hyphenation.
non-preinstalled	Do not use. Use "not preinstalled."
non self-identifying (noun, modifier)	Note hyphenation.
not preinstalled	Use instead of "non-preinstalled."

TABLE B–1 Term Usage and Style *(Continued)*

Term	Usage
now	Use only in a document that you know will be updated regularly, for example, in release notes. You can also use a specific date, for example, "August 2009."
O	
object ID	Do not use. Use "OID."
object identifier (OID)	Note that the abbreviation is "OID," not " object ID."
object-oriented	modifier
off and on	If referring to a switch on a specific piece of equipment, use the capitalization that appears on the equipment (usually ON and OFF). If referring to a switch that is marked with symbols only, use initial capitalization and include the symbol in parentheses, for example, "On (|)" or "Off (o)." Otherwise, use lowercase letters, for example, if you are telling the user to turn the power on or off.
offline	adverb, modifier
off-load	verb
ohmmeter	noun
OID	Abbreviation for "object identifier." Note that the abbreviation is not "object ID."
OK button	Note capitalization.
`ok` prompt	Note lack of capitalization and monospace font.
on-board	modifier
on-demand (modifier)	Note hyphenation.
online	adverb, modifier
online help	When referring to online help, use instead of "help" or "Help."
onscreen	modifier
open source (noun), open-source (modifier)	Note lack of capitalization. Hyphenate when used as a modifier.
optical disc	Note spelling.
option, optional component	Use instead of "x-option" or "X-option."
output	noun (not a verb)
overtemperature (modifier)	Note lack of hyphenation.

TABLE B–1 Term Usage and Style (*Continued*)

Term	Usage
oz	Abbreviation for "ounce."
P	
p	Abbreviation for "pico" (prefix).
panhead screw	noun
parallel database (PDB)	Note lack of capitalization of spelled-out form. Note spelling of "PDB."
parallel optical link (PAROLI)	Note lack of capitalization of spelled-out form. Note spelling of "PAROLI."
PAROLI	Acronym for "parallel optical link." Note spelling of "PAROLI."
part number, part no.	Note lack of capitalization. Note period in abbreviated form. Do not abbreviate as "PN" or "P/N."
path name (noun), path-name (modifier), *pathname* (variable)	Hyphenate when used as a modifier.
PC file viewer	Note capitalization.
PCI card	Note capitalization.
PC launcher	Note capitalization.
PDB	Abbreviation for "parallel database." Note spelling of "PDB."
PDF	Abbreviation for "Portable Document Format." Do not use the spelled-out form for most technical audiences.
percent	Use "percent" in text. Use the symbol % in tables and slides to conserve space.
permit	See "allow."
Personality ASIC	Note capitalization.
phase-locked loop (PLL)	Note hyphenation and lack of capitalization of spelled-out form.
Phillips screw, Phillips screwdriver, No. 2 Phillips screwdriver	Do not use the "-head" suffix, for example, "Phillips-head screw."
Photo CD format	noun
pinout	noun
`pkgadd`	Use "`pkgadd` utility" or "`pkgadd` script," not "`pkgadd` file." Note monospace font.
please	Do not use. You are not making a request, you are telling the reader to do something.

TABLE B–1 Term Usage and Style (Continued)

Term	Usage
PLL	Abbreviation for "phase-locked loop."
plug (noun, verb)	Do not use the noun form. Use "connector" instead. For the verb form, use "connect" instead, *except* in the context of "plug and play."
plug and play	Use to indicate the general ability to connect pieces of hardware together so they "just work." For example, "The goal of the Solaris device configuration framework is for devices to plug and play, with no user configuration required."
Plug and Play (PnP)	Use to refer to devices that conform to published Plug and Play specifications. For example, "Non-Plug and Play devices are referred to as 'legacy devices.'"
plug-in (noun, modifier)	Note hyphenation.
p.m.	Abbreviation for "post meridiem (after noon)." Note the periods.
PN, P/N	Do not use. Use "part number" or "part no." instead.
PnP	Abbreviation for "Plug and Play." Note capitalization.
pop-up	modifier
port A	Note capitalization.
postinstall (verb)	Write "install" unless you are describing a process that literally takes place directly after installation.
postinstallation (noun, modifier)	Note lack of hyphenation.
postpend	Do not use. Use "append."
power cycle (noun, verb), power-cycle (modifier)	Hyphenate when used as a modifier.
power down (verb), power-down (modifier)	Hyphenate when used as a modifier. Powering down is a procedure. If you are writing about flipping a switch, use "Turn the power off."
power grid	Note two words.
power off (verb), power-off (modifier)	Hyphenate when used as a modifier.
power on (verb), power-on (modifer)	Hyphenate when used as a modifier.
Power On/Off switch	Note capitalization and the slash.
power up (verb), power-up (modifier)	Hyphenate when used as a modifier. Powering up is a procedure. If you are writing about flipping a switch, use "Turn the power on."

TABLE B–1 Term Usage and Style *(Continued)*

Term	Usage
preceding	Use instead of "above" or "before" to refer to the location of an immediately prior piece of information. Consider using "previous" or referring to the specific section heading or figure number, for example.
preinstall (verb)	Use "install" unless you are describing a process that literally takes place before installation.
preinstallation (noun, modifier)	Note lack of hyphenation.
prepend (verb)	Use to denote the action of placing data at the beginning. For example, "The field also includes a prefix attribute that is prepended to each entry in the output zip file."
preprocessor	noun
presently	Use only in a document that you know will be updated regularly, for example, in release notes. You can also use a specific date, for example, "August 2009."
press	When providing a keyboard or mouse button instruction, use instead of "depress," "hit," or "type." For example, "Press Enter."
preventative	Do not use. Use "preventive."
preventive	Use instead of "preventative."
previous	Use instead of "above" or "before" to refer to the location of another piece of information mentioned earlier. Consider using "preceding" or referring to the specific section heading or figure number, for example.
printout	noun
pseudo-device	noun, modifier
pull-down	modifier
pull-right	modifier
Q	
Quality of Service (QoS)	Note capitalization.
queuing	Note spelling.
R	
rackmount	noun, modifier, verb
rack unit, rack units (U)	Note lack of capitalization of spelled-out form. The abbreviation "U" is singular, uppercase, with no space between the abbreviation and the numeral, for example, "2U." Abbreviation is *not* "RU."

TABLE B–1 Term Usage and Style *(Continued)*

Term	Usage
RAID 5, RAID Level 5 (noun); RAID-5 (modifier)	Note capitalization. Hyphenate when used as a modifier.
RAS	Acronym for "reliability, availability, and serviceability."
raster file (noun), raster-file (modifier)	Hyphenate when used as a modifier.
Ready signal	Note capitalization.
real mode (noun), realmode (modifier)	Note lack of hyphenation.
real time (noun), real-time (modifier)	Hyphenate when used as a modifier.
rear panel	Do not use. Use "back panel."
reattach	verb
recommend	Do not use. Just go ahead and recommend. For example, write "Back up all your files once each week," not "It is recommended that you back up all your files once each week." In some less definitive instances, describe the circumstances in which backing up is recommended. For example, "Back up your `.login` file whenever you modify it."
re-create	verb
reinsert	verb
reinstall	verb
reinstallation	noun
reliability, availability, and serviceability (RAS)	Note lack of capitalization of spelled-out form.
Remote Method Invocation (RMI)	Note capitalization.
removable	Note spelling.
remove screw	Use for noncaptive screws. Do not use "unscrew" or "screw."
replaceable	Note spelling.
reprobe	verb
restore (verb, modifier)	Do not use as a noun.
retry	verb
revolutions per minute (rpm)	Note lack of capitalization.

TABLE B–1 Term Usage and Style *(Continued)*

Term	Usage
`.rhost` file	Pronounced "dot r-host file." The preceding indefinite article should be "a," not "an."
right-click	verb
right-justified	modifier
right-hand side	Do not use. Use "right side."
right side	Use instead of "right-hand side."
right-to-use (RTU) license	Note hyphenation and lack of capitalization of spelled-out form.
RJ-45	noun, modifier
RMI	Abbreviation for "Remote Method Invocation."
root	Use to refer to the root (/) directory or root (/) file system. Do not use this term by itself. Use, for example, "root user," "root directory," or "root (/) file system."
rpm	Abbreviation for "revolutions per minute." Note lack of capitalization.
RS-232-C, RS-423…	noun, modifier
RTU license	Abbreviation for "right-to-use license."
RU	Do not use. See "U."
run level	noun, modifier
run-on	modifier
runtime	noun, modifier
S	
s, sec	Abbreviation, acronym for the time unit of measurement "second."
sample (noun, modifier, verb)	Do not use as a noun or modifier in reference to code examples. Use "example" (noun, modifier) instead.
SBus	Note capitalization.
scalable	Note spelling.
Scalable Shared Memory (SSM)	Note capitalization.
screen capture	Use instead of "screen shot."
screen shot	Do not use. Use "screen capture."
screw (noun, verb)	Do not use as a verb for captive screws. Use "loosen" or "tighten."

TABLE B–1 Term Usage and Style *(Continued)*

Term	Usage
scrollbar	noun
SCSI	Pronounced "scuzzy" so the preceding indefinite article should be "a," not "an."
SCSI-2	modifier
SCSI address 0	Note capitalization.
SCSI bus	Note capitalization.
SCSI device identifier (SCSI ID)	Note capitalization. Note spelling of "SCSI ID."
search results	Use "search results" to refer to information obtained when searching the web, for example, unless doing so conflicts with a user interface convention. See also "hits."
Secure Sockets Layer (SSL)	Note capitalization.
SEEPROM	Acronym for "serially electrically erasable PROM." Do not spell out "PROM."
self-test	noun, modifier
semiconductor	noun, modifier
serially electrically erasable PROM (SEEPROM)	Note capitalization. Do not spell out "PROM."
serial port A, serial port B...	Note capitalization. You can also use the terms "COM1, COM2" and so on.
server	Because a network includes many types of servers, assign a limiting modifier and use it throughout, for example, "print server" or "session server." Use also when talking about the relationship with a client.
set-top box (STB)	Note hyphenation and lack of capitalization of the spelled-out form.
setup (noun, modifier), set up (verb)	Note lack of hyphenation.
shutdown (noun, modifier), shut down (verb)	Note lack of hyphenation.
side rail	noun
simple	Avoid. This term usually does not apply to technical information.
simply	Avoid. This term usually does not apply to technical information.
single-density	modifier

TABLE B–1 Term Usage and Style *(Continued)*

Term	Usage
single point of failure (SPOF) (noun), single-point-of-failure (SPOF) (modifier)	Note lack of capitalization of spelled-out form. Hyphenate when used as a modifier.
single-sided	modifier
single-tasking	modifier
slave I/O controller	Note capitalization and the slash.
smart card (noun, modifier)	Lowercase when used generically.
SMP (modifier)	Abbreviation for "symmetrical multiprocessing."
spacebar	Note one word and lack of capitalization.
SPOF	Abbreviation for "single point of failure" (noun) or "single-point-of-failure" (modifier).
spreadsheet	noun, modifier
sq	Abbreviation for "square."
SSL	Abbreviation for "Secure Sockets Layer."
SSM	Abbreviation for "Scalable Shared Memory."
SSP	Abbreviation for "System Service Processor."
stand-alone (modifier)	Note hyphenation.
start (noun, verb)	When used as a verb, use instead of "start up" or "launch."
startup	noun, modifier
STB	Abbreviation for "set-top box."
stream, Stream, STREAMS	Use "stream" to refer to a data stream.
	Use "Stream" to refer to a communication Stream that is part of the larger STREAMS mechanism.
	Use "STREAMS" to refer to the mechanism that defines interface standards for character I/O within the kernel and between the kernel and user level.
Super I/O	Note two words, capitalization, and the slash.
supersede	Note spelling.
superuser	Note one word. Equivalent to "root user" or, in role-based access control (RBAC), "primary administrator." Preferred term: in general usage, "superuser."

TABLE B–1 Term Usage and Style *(Continued)*

Term	Usage
Supplement CD	Use instead of "Update CD."
surf	Do not use when you mean "navigate" the web.
symmetrical multiprocessing (SMP) (modifier)	Note lack of capitalization and hyphenation of spelled-out form.
system	When possible, use a more descriptive term, such as "mail server" or "remote system."
System Binary Interface	Note capitalization, which is an exception to interface naming conventions.
system name (noun), *system-name* (variable)	Note hyphenation of variable.
System Service Processor (SSP)	Note capitalization.
T	
target ID *n*	Note capitalization. The last character represents a variable.
Tbyte, TB in tables only	Acronym, abbreviation for "terabyte." In computer terminology, this represents 2^{40}, or roughly a thousand gigabytes.
Telnet window	Note capitalization.
text-only	modifier
that	This word is often misused in nonrestrictive clauses or phrases instead of "which." Use "that" for restrictive clauses or phrases. For example, "I like mysteries that are suspenseful." (I like only those mysteries that are suspenseful.) "I like mysteries, which are suspenseful." (I like all mysteries, and mysteries have the attribute of being suspenseful.)
there are	Because "there are" is ambiguous, avoid using this phrase at the beginning of a sentence or clause.
there is	Because "there is" is ambiguous, avoid using this phrase at the beginning of a sentence or clause.
thick Ethernet	Note capitalization.
Thicknet	Note capitalization.
Thinnet	Note capitalization.
third party (noun), third-party (modifier)	Hyphenate when used as a modifier.

TABLE B–1 Term Usage and Style *(Continued)*

Term	Usage
tighten or loosen noncaptive screws	Use instead of "screw" or "unscrew."
time out (verb), timeout (noun, modifier)	Note lack of hyphenation.
timesharing (noun, modifier)	Note lack of hyphenation.
time stamp (noun), time-stamp (modifier, verb)	Hyphenate when used as a modifier or verb.
time zone (noun)	Note two words.
TIP connection	Note capitalization.
TOD	Abbreviation for "time-of-day" (modifier).
token reader	noun
Token Ring	Use initial capitalization only when referring specifically to the IBM Token Ring or the IBM Token-Ring Network.
toolkit	noun
toward	Note lack of final "s."
TPE Link Test	Note capitalization.
trackball	noun
trade-off (noun, modifier), tradeoff (verb)	Hyphenate when used as a noun and modifier. Use verb form sparingly.
transition	noun (not a verb)
translation-lookaside	noun, modifier
triple-height	modifier
trivial	Do not use. A cliche, this word does not mean "easy." It means "insignificant."
troubleshooting (noun, modifier), troubleshoot (verb)	Note change for verb.
true-color	modifier
twisted pair (noun), twisted-pair (modifier)	Hyphenate when used as a modifier.
Twisted-Pair Ethernet Link Test	Note capitalization.

TABLE B–1 Term Usage and Style *(Continued)*

Term	Usage
type (noun, verb)	When used as a verb, use instead of "enter" if the action involves the actual typing of data. Also, use instead of "enter" to avoid confusion with the Enter key, which you "press."
Type-5	modifier
U	
U	Abbreviation for "rack unit" and "rack units." The abbreviation "U" is singular, uppercase, with no space between the abbreviation and the numeral, for example, "2U." Abbreviation is *not* "RU."
UDF	Abbreviation for "Universal Disk Format."
UL	Abbreviation for "Underwriters Laboratories."
UltraSCSI (noun, modifier)	Note one word, capitalization, and lack of hyphenation.
Ultra Wide SCSI (noun, modifier)	Note capitalization and lack of hyphenation.
unconfigure (verb)	Use instead of "deconfigure."
Underwriters Laboratories (UL)	Note capitalization. Note no apostrophe in "Underwriters."
Universal Disk Format (UDF)	Note capitalization.
Universal Serial Bus (USB)	Note capitalization of spelled-out form, which is an exception to bus naming conventions.
unscrew	Do not use. Use "loosen" or "remove."
unshielded twisted-pair (UTP)	Note hyphenation and lack of capitalization of spelled-out form.
uninstall (verb)	Not a noun or modifier. Use instead of "deinstall."
uninstallation	Do not use.
uniprocessor	noun
UNIX	Note capitalization.
Update CD	Do not use. Use "Supplement CD."
upgradable	Note spelling.
upload	verb, modifier
uppercase	noun, modifier, verb
upper left (modifier)	Use instead of "upper left-hand."
upper right (modifier)	Use instead of "upper right-hand."
uptime	noun

TABLE B–1 Term Usage and Style *(Continued)*

Term	Usage
USB	Abbreviation for "Universal Serial Bus."
U.S., U.S.A.	Abbreviation for "United States" and for "United States of America." These abbreviations include periods.
usable	Note spelling.
user-defined	modifier
user-friendly	modifier
userland (noun, modifier)	Means "not in the kernel." Use sparingly.
user name (noun), *username* (variable)	Note lack of hyphenation.
using	Be careful. "Configure the roles using a single rights profile" can be interpreted either as "Use a single rights profile to configure the roles" or "Configure the roles that are using a single rights profile." If you mean to use the verb form, make sure that you supply the required preposition: "Configure the roles *by* using a single rights profile."
UTC	Abbreviation for "Coordinated Universal Time." Note spelling of "UTC."
UTP (noun, modifier)	Abbreviation for "unshielded twisted-pair."
V	
V	Abbreviation for "volt."
variable-length	modifier
Veritas File System (VxFS)	Note capitalization. Note spelling of "VxFS."
versus, vs.	Do not use. Use "compared with."
vertexes	Plural of vertex. Use instead of "vertices."
via	Do not use. Use the more common equivalents "through," "by means of," "using," or "by way of."
vice versa	Do not use. Use "conversely" or "with the order reversed."
video disc	Note spelling.
video display device	Use instead of "graphics card."
`vi` editor	Note lack of capitalization and monospace font.
VIP	Acronym for "virtual IP address." Note spelling of "VIP."
virtual IP address (VIP)	Note capitalization. Note spelling of "VIP."

TABLE B–1 Term Usage and Style *(Continued)*

Term	Usage
virtual LUN (VLUN)	Note capitalization.
VMEbus	Note capitalization.
volt-ohmmeter	noun
volume management	Use "volume management" when referring to the technology that manages removable media. For example, "The volume management daemon, `vold`, automatically mounts and unmounts removable media such as Zip or Jaz drives."
VxFS	Abbreviation for "Veritas File System." Note capitalization. Note spelling of "VxFS."
W	
W	Abbreviation for "watt."
W3C	Abbreviation for "World Wide Web Consortium." Use "W3C" sparingly.
wake-on-lan cable	Note hyphenation and lack of capitalization.
want	Use instead of "desire" or "wish."
web	Use lowercase except in the term "World Wide Web" or in specific product names.
web browser	Do not use. Use "browser."
webmaster	Note one word.
web server	Note two words and lack of capitalization of this generic term.
web site	Note two words.
Wh	Abbreviation for "watt-hour."
when	Use to refer to an inevitable event. See also "if."
whether or not	The words "or not" are usually unnecessary. This whole phrase can sometimes be replaced with "if," so make sure you are using the correct conditional term.
which	This word is often misused in restrictive clauses or phrases instead of "that." Use "which" for nonrestrictive clauses or phrases. For example, "I like mysteries that are suspenseful." (I like only those mysteries that are suspenseful.) "I like mysteries, which are suspenseful." (I like all mysteries, and mysteries have the attribute of being suspenseful.)

TABLE B–1 Term Usage and Style *(Continued)*

Term	Usage
white paper	Note two words.
white space	Note two words.
Wide SCSI	modifier
wildcard	noun, modifier
wireless LAN (WLAN)	Note capitalization.
wish	Do not use. Use "want."
workaround (noun), work around (verb)	Note lack of hyphenation.
workflow (noun, modifier)	Note lack of hyphenation.
workgroup	noun, modifier
workload	noun
worksheet	noun
workspace	noun, modifier
Workspace menu	Note capitalization.
workstation	Do not use unless you are referring to a specific piece of hardware.
World Wide Name (WWN)	Note capitalization.
World Wide Web (WWW)	Note capitalization.
World Wide Web Consortium (W3C)	Note capitalization. Use abbreviation "W3C" sparingly.
wraparound (noun, modifier), wrap around (verb)	Note lack of hyphenation.
write-back	modifier
write-enable (verb), write-enabled (modifier)	Note hyphenation. Note "d" at end of modifier.
write-protect (verb), write-protected (modifier)	Note hyphenation. Note change for modifier.
WWN	Abbreviation for "World Wide Name."
WWW	Abbreviation for "World Wide Web."
X	
x86	Note lack of capitalization.

TABLE B–1 Term Usage and Style *(Continued)*

Term	Usage
x-axis	Note hyphenation and lack of capitalization.
XBus	Note capitalization.
XDBus	Note capitalization.
XIR	Abbreviation for "externally initiated reset." Note spelling of "XIR."
XON/XOFF	Note capitalization and the slash.
x-option, X-option	Do not use. Use "option" or "optional component" depending on the context.
Y	
y-axis	Note hyphenation and lack of capitalization.
Z	
z-axis	Note hyphenation and lack of capitalization.
z-buffer	Note hyphenation and lack of capitalization.
zeros	Note spelling.

APPENDIX C

Typographic Conventions

Typographic conventions help a reader distinguish special uses of fonts. This appendix shows elements used in typical technical documentation for which typographic conventions are established.

To provide consistency in typographic cues for readers, use the following conventions in your document.

Note – The examples in the following table show typographic conventions only. These examples are not meant to dictate other aspects of formatting technical information such as how code should be indented or where braces should be placed in code examples.

TABLE C–1 Typographic Conventions

Text[1]	Specification[2]	Examples
Alert names	Lowercase, regular text font	critical alert, major alert, minor alert
Arguments	Monospace font	`-f`, `Slapi_PBlock`
Arrow keys	Lowercase, regular text font	Press the up arrow, then the left arrow.
ASIC names	Lowercase for name, regular text font	address multiplexer ASIC
		reset, interrupt, scan, and clock (RISC) ASIC
		data arbiter ASIC
Board names, when referring to a specific model of board	Initial capitals, regular text font	GT SBus Adapter Board
		600MP System Board
		600MP Expansion Memory Board

[1] A plural term is usually generic and in those cases should not appear in monospace font, for example, makefiles.

[2] The point size of the font for monospace elements should match the point size of the text within which they appear, including a heading, paragraph, table, or caption.

TABLE C–1 Typographic Conventions (Continued)

Text[1]	Specification[2]	Examples
Board types, when referring to a generic type of board	Lowercase, regular text font	auxiliary video board, system board, power distribution board
Book titles	Italic, regular text font	See the *PlirgPak Reference Guide.*
Bus names	All capitals for the bus name, but not for the word "bus"; regular text font	AFX bus, PROM bus, PCI bus
Bus types	Lowercase, regular text font	address bus, host bus
Button names	Initial capitals for the button name, but not for the word "button"; regular text font	Power button
Card names	Initial capitals or all capitals for the card name, but not for the word "card"; regular text font	Control card, TNT card
Card types	Lowercase, regular text font	optics card, smart card, graphics card
CD titles	Initial capitals, regular text font	PlirgSoft Installation CD
Chapter titles or section headings	Initial capitals in quotation marks, regular text font	See Chapter 1, "Getting Started."
Classes	Monospace font	`Morf`, `AncestorEvent`, `DoubleBuffer`

Code examples	Monospace font	`struct inode {`
		`struct inode *i_chain[2];`
		`struct vnode i_vnode;`
		`struct vnode *i_devvp;u_short i_flag;`
		`};`

Command field names	Monospace font	Look at the `Flags` field in the output.
Command names or options	Monospace font	Type `ls -a` to list all files.
Constants	Monospace font	`NULL`, `NIS_CBError`
Daemon names	Monospace font	`cron`, `lpsched`
Database files	Monospace font	`bootparams.dir`, `hosts.byaddr.dir`
Database names	Initial capitals, regular text font	Clientbase

[1] A plural term is usually generic and in those cases should not appear in monospace font, for example, makefiles.

[2] The point size of the font for monospace elements should match the point size of the text within which they appear, including a heading, paragraph, table, or caption.

TABLE C–1 Typographic Conventions (Continued)

Text[1]	Specification[2]	Examples
Data types	Monospace font	`short`, `boolean`, `kcondvar_t`
Directory names or path names	Monospace font	Check the book directory in `/docs/work`.
DOS commands	All capitals, monospace font	`RENAME`, `FDISK`
Driver names	Monospace font	`asy`, `ata`, `dnet`
Email addresses	Monospace font	Send questions to `support@plirgware.com`.
Emphasized words or new terms[3]	Italic, regular text font	You *must* delete your old files. A *cache* is a copy that is stored locally.
Environment variables	All capitals, monospace font	`PATH`, `PROMPT`, `TMPDIR`, `TERM`
Equations	Italic, regular text font for replaceable items, and monospace font for constants and mathematical operators	*availability* = *update* `/` `(`*uptime* + *downtime*`)` × `100%`
Error messages, when you know the exact wording	Monospace font	`stty: No such device or address.`
Event names	Monospace font for the event name, but lowercase, regular text font for the word "event"	`snmp-mibII.system` event
Exceptions	Monospace font	`IllegalStateException`, `NullPointerException`
Field names in a programming context	Monospace font	`NULL_ATTRIBUTE_VALUE`, `au_sample_rate`
Field names from a graphical user interface in text	No colon or ellipsis points, regular text font	[Field name appears as Tag Type: ___________] Type the name of the tag in the Tag Type field.
File format names	Monospace font	`mif`, `mpeg`, `java`
File names, generic	Regular text font	When you use makefiles, additional swap space might be required.
File names, specific	Monospace font	Delete all `.bak` and `core` files.

[1] A plural term is usually generic and in those cases should not appear in monospace font, for example, makefiles.

[2] The point size of the font for monospace elements should match the point size of the text within which they appear, including a heading, paragraph, table, or caption.

[3] In some single-source authoring environments, emphasized text is shown in italic in print but displayed in bold online.

TABLE C–1 Typographic Conventions *(Continued)*

Text[1]	Specification[2]	Examples
File permissions	Monospace font	To make the file readable and writable by the file owner and readable by everyone else, change the permissions to u+x.
File system names	Monospace font	/usr, /opt, /var
File system types in text	All capitals, regular text font	UFS, NFS, HSFS
Flags	Monospace font	The flag THR_NEW_LWP is passed to the function thr_create() to create LWP.
Functions, function calls, and interfaces	Monospace font	abs(), ctermid()
Graphical user interface elements	Regular text font	Choose Save from the File menu. Click OK.
Headers	Monospace font	exception, exception.h
HTML tags	Monospace font	<h1>, <body>
Interface names in text, when referring to a specific type of interface	Initial capitals for the interface name, but most often not for the word "interface"; regular text font	Ethernet interface, InterDomain Network interface, Windows interface, SCSI interface
Interfaces (in a programming context)	Monospace font	Runnable, la_objfilter(), PathIterator
Interface types, when referring to a broad category of interface	Lowercase, regular text font	user interface, network interface, serial interface, command-line interface
IP addresses	Monospace font	10.255.255.255
Key names[4]	Initial capitals, regular text font	Press the Control key.
Keywords in files (see also "reserved words")	Monospace font	netmask
LED names	Initial capitals for the LED name, but all capitals for the word "LED"; regular text font	Warning LED
Machine names	Monospace font	newstop, dickens
Macros	All capitals, monospace font	MAX(), MIN()

[1] A plural term is usually generic and in those cases should not appear in monospace font, for example, makefiles.

[2] The point size of the font for monospace elements should match the point size of the text within which they appear, including a heading, paragraph, table, or caption.

[4] See "Key Name Conventions" on page 84.

TABLE C–1 Typographic Conventions *(Continued)*

Text[1]	Specification [2]	Examples
Man page names	Monospace font for the man page name only, regular text font for the parentheses and section name	`boot`(1M), `connect`(3SOCKET)
Menu options, menu names	Initial capitals, no colon or ellipsis points, regular text font	Choose Go To from the Page menu. [Option appears as Insert Markup on screen.] Choose Insert Markup from the Markup menu.
Method and constructor names[5]	Monospace font	`add`, `add(Object)`, `add(int,Object)`
Mode names	Initial capitals for the mode name, but not for the word "mode"; regular text font	Config mode, Edit mode, Run mode
Module names	Initial capitals for the module name, but not for the word "module"; regular text font	Wireless Connection Wizard module, Mod module
Module types, when referring to a generic type of module	Lowercase, regular text font	locking module, software module, stacking module, web module
Node names	Initial capitals for the node name, but not for the word "node"; regular text font	Control node, Packet Data Serving node
Node types, when referring to a generic type of node	Lowercase, regular text font	junction node, interface node, application server node
Operators and operations	Monospace font	`&&`, `!=`, `instanceof`
Package names, installation	Monospace font	`PLRGman`, `PLRGlib`
Package names, Java	Monospace font	`java.io`, `java.security.cert`
Parameters	Monospace font	`foo(x, y)` (`x` and `y` are parameters for `foo()`)
Partition names, the PC disk partition `fdisk`	Monospace font	Reformat the `PLIRGPAK` partition.
Patch IDs (referred to as "update IDs" by some groups)	Regular text font	111879-01

[1] A plural term is usually generic and in those cases should not appear in monospace font, for example, makefiles.

[2] The point size of the font for monospace elements should match the point size of the text within which they appear, including a heading, paragraph, table, or caption.

[5] Omit parentheses for the general form of methods and constructors. When a method or constructor has multiple forms and you mean to refer to a specific form, use parentheses and argument types.

TABLE C–1 Typographic Conventions *(Continued)*

Text[1]	Specification[2]	Examples
Permission names	Lowercase, regular text font	execute permission, read permission, write permission, read/write permissions
Port names in text	All capitals, regular text font	COM1, LPT3
Print services	Regular text font	LP print services
Printer names	Monospace font	Assign `catalpa` as the printer name.

Prompts	Monospace font, unless from a GUI	%, $, #
		`system123%`
		`system123$`
		`system123#`
		`>`
		`ok`

Property names	Monospace font	Specify the `nohangup` property.
Protocol names	Initial capitals including the word "Protocol"; regular text font	File Transfer Protocol
README files, generic	All capitals, regular text font	See the README file in `/tmp`.
README files, specific file name	Monospace font	See the `README.html` file.
Release media (CD, diskette, and DVD titles)	Initial capitals, regular text font	Insert the first PlirgSoft 1.1 Update CD.
Reserved words, or keywords (see also "keywords in files")	Monospace font	`if`, `else`, `int`, `float`
Return values	Monospace font	The function `open_max()` returns `TRUE`. A value of `-1` is returned.
Root symbol (slash)	Monospace font	Go to the `/` (root) directory.
Routines	Monospace font	The `time()` routine computes time.
Run levels	Monospace font	Type `init S` to change to run level `S`.
Settings in windows	Initial capitals, regular text font	Check the Pair Kern setting.
Script names	Monospace font	`check`, `rc`, `sbin/rc3`

[1] A plural term is usually generic and in those cases should not appear in monospace font, for example, makefiles.

[2] The point size of the font for monospace elements should match the point size of the text within which they appear, including a heading, paragraph, table, or caption.

TABLE C–1 Typographic Conventions *(Continued)*

Text[1]	Specification [2]	Examples
Signal names	Initial capitals for the signal name, but not for the word "signal"; regular text font	Clear To Send signal
Slice names	Monospace font	`s0`, `/opt`
Slot names	Initial capitals for the name or letter of the slot, but not for the word "slot"; regular text font	slot 0, slot A
Software elements that need to be distinguished from regular text	Monospace font	`file` realm, `enabled` state
Structures[6]	Monospace font	`proc`, `sockaddr`
Switch names and switch settings	Initial capitals for the switch name, but not for the word "switch"; regular text font	Power switch
System calls or subroutines	Monospace font	`read()`, `open()`, `_lwp_info()`
System error codes	All capitals, monospace font	`EINVAL`, `ENOENT`, `setErr()`
System names	Monospace font	Move the file to the `nesfal` system.
System variables	Monospace font	The system variable in `set maxusers=40` is `maxusers`.
Test names	Initial capitals for the test name, but not for the word "test"; regular text font	Alarm Card test
Tool or utility names	Case-sensitive, monospace font if the command name is used; initial capitals, regular text font if the formal name (if any) is used	`grep`, `sed`, `vi`, `appletviewer` Formal names: Nametool, Mail Tool, Bugfix
URLs or domain names	Monospace font	`http://www.plirg.com` `example.com`, `my.site.com`

[1] A plural term is usually generic and in those cases should not appear in monospace font, for example, makefiles.

[2] The point size of the font for monospace elements should match the point size of the text within which they appear, including a heading, paragraph, table, or caption.

[6] The file index structures inode, vnode, and rnode are not literals and use the default body font.

TABLE C–1 Typographic Conventions *(Continued)*

Text[1]	Specification[2]	Examples
User input in running text or steps	Monospace font	Use `ls -a` to list all files. **1. Type** `teh` **in the Find text field.** **Note:** Some authoring tools might not allow regular-weight alternate font material within a bold step.
User input contrasted with computer output or with a prompt in code examples	Bold, monospace font	`system123%` **`su`** `Password:` **`# tar -xvf`** *filename*
User names and IDs	Monospace font	User `davemc` has a user ID of `1001`. User `root` destroyed the system.
Variables, command-line placeholders	Italic, regular text font	Name the file *filename* `TOC.doc`. Go to `/`*hostname*`/docs/work`. `%` **`chmod 600`** *filename*
Web site names	Monospace font	For more details, go to `http://www.plirg.com`.
Window button names	Initial capitals, regular text font	Click Apply to save your changes.
Window elements	Regular text font	In the Alarms pane, click Remove.

[1] A plural term is usually generic and in those cases should not appear in monospace font, for example, makefiles.

[2] The point size of the font for monospace elements should match the point size of the text within which they appear, including a heading, paragraph, table, or caption.

Checklists and Forms

This appendix provides samples of common publications forms and checklists. You might need to modify them to reflect your own company processes.

The following checklists and forms appear in this appendix:

Manuscript Tracking Chart

Document Title: ____________________

Project: ____________________

Writer: ____________________ No. of pages: __________

Edit Requested	Sent to Editor (1st review)	Returned to Writer (1st review)	Sent to Editor (2nd review)	Returned to Writer (2nd review)	Notes
Alpha					
Develop. edit					
Copy edit					
Proofread					
Release check					
Beta					
Develop. edit					
Copy edit					
Proofread					
Release check					
FCS					
Develop. edit					
Copy edit					
Proofread					
Release check					

Checked: ❑ Front matter ❑ Back matter ❑ Cross-references ❑ Spelling

Dates of files to production:

Alpha __________ Beta __________ FCS __________ Print __________

Request for Editing Form

Document Title: ____________________

Writer: ____________________ Product code name: __________

Phone no.: __________ Email address: ____________________

Number of pages: ________ Date submitted: ________ Return by: ________

Target audience: ____________________

Type of edit requested: ❑ Developmental ❑ Copy ❑ Proofread

Development stage: ❑ Alpha ❑ Beta ❑ FCS

❑ Other:

Is this document part of a set? ❑ Yes ❑ No

Name of set: ____________________

Has this document been edited before? ❑ Yes ❑ No

Editor: ____________________ Date of prior edit: ________

Type of edit: ____________________

Were comments incorporated into this draft? ❑ Yes ❑ No

Do specific sections need particular attention? ❑ Yes ❑ No

If so, which ones? ____________________

Does the set have a style sheet? ❑ Yes ❑ No

If so, is it included for the editor? ❑ Yes ❑ No

Check for the following items before giving a book to the editor:

❑ Title page, credits page, TOC, LOF, LOT, and preface are current.

❑ Glossary and index are complete and current.

❑ Page numbers and footers are correct.

❑ There are no unexplained blank pages.

❑ Graphics are incorporated or content and placement are indicated.

❑ Cross-references are updated.

❑ Trademarked terms are marked appropriately on first reference in text.

❑ Spelling checker was run on all files.

Comments? ____________________

____

____

Artwork Request Form

Full document title: ______________________________

Full document part number: ______________________________

Requestor: ____________________ Phone no.: ______________

Date due to writer (art delivered for proofing by midnight): ______________

Document is: ❑ Beta ❑ FCS ❑ Other

Document created in: ❑ Epic ❑ Frame ❑ Other

No.	Figure caption to be used in this document (and used in previous document, if different)	Existing Control No.	New/Rev. Control No.
			Column for Illustrator Use Only

Technical Review Cover Letter

TO: *Reviewer List*

FROM: *Writer Name*

SUBJECT: Technical Review of *Name of Document*

DATE: *Date*

The attached manuscript is the technical review version of *Name of Document*. Please review the entire manuscript, paying special attention to the notes and questions to reviewers. All open issues and unanswered questions must be resolved for this technical review to be complete.

I would appreciate your general comments as well as specific answers to the issues raised in the notes. Please give detailed and thorough responses. Also, please address the specific review responsibilities of your department.

[*Add any comments regarding specific issues or content.*]

Your review must be returned by 5 p.m. on *Date*. After reviewing your responses, I will discuss any discrepancies at the technical review sign-off meeting, which will be held on *Date* at *Time* in *Name* Conference Room.

Due to the tight production schedule, please make wording or style suggestions only if they affect the technical accuracy of the text.

Thank you very much for your attention to this document. Your comments are appreciated and contribute greatly to improving the quality of *Company Name* documentation.

Print Specification

Date:
Contact name
Contact phone no.

Company name
Company address
Company address
Fax number

Printing specification for: ______________________
Product description: ______________________
Documents to vendor (date): ______________________
Product ship date: ______________________
Quantity to print: ______________________

Manuals: (If the documentation set is small, fill in the table below. For larger jobs, attach a list of manuals, page counts, and the preferred binding method.)

Part Number	Title	Page Count	Binding

Format size: ______________________________

Text stock: ______________________________

Cover stock: ______________________________

Cover art: ______________________________

Tabs: ______________________________

Cards: ______________________________

Labels: ______________________________

Special boxes or cartons: ______________________________

Media: ______________________________

Printing process: ______________________________

Proofing requirements: ______________________________

____

Assembly instructions: ______________________________

____

General comments: ______________________________

____

____

Please provide ____________ with __ check copies at time of first customer ship.

Return all original hand-off material to ________________ at completion of job.

APPENDIX E

Recommended Reading

This appendix lists resources targeting the needs of technical writers, editors, production staff, and illustrators.

Note – This list is for your reference only. You must seek permission from the publisher owning the copyrighted source material before reprinting any text that you select, either in book or electronic form.

This appendix covers the following categories:

Desktop Publishing and Document Design

Graham, Lisa. *Basics of Design: Layout and Typography for Beginners.* 2d ed. Clifton, N.Y.: Delmar Learning, 2005.

Lichty, Tom. *Design Principles for Desktop Publishers.* 2d ed. Belmont, Calif.: Wadsworth Publishing Co., 1994.

Schriver, Karen A. *Dynamics in Document Design: Creating Text for Readers.* New York: John Wiley & Sons, 1994.

Williams, Robin, and John Tollett. *Robin Williams Design Workshop.* 2d ed. Berkeley, Calif.: Peachpit Press, 2006.

Editing Standards

Einsohn, Amy. *Copyeditor's Handbook: A Guide for Book Publishing and Corporate Communications.* 2d ed. Berkeley, Calif.: University of California Press, 2005.

Gordon, Karen Elizabeth. *The Deluxe Transitive Vampire: The Ultimate Handbook of Grammar for the Innocent, the Eager, and the Doomed.* 1st ed. New York: Pantheon Books, 1993.

Gordon, Karen Elizabeth. *The New Well-Tempered Sentence: A Punctuation Handbook for the Innocent, the Eager, and the Doomed.* Revised and expanded. New York: Ticknor & Fields, 1993.

Judd, Karen. *Copyediting: A Practical Guide.* 3d ed. Los Altos, Calif.: Crisp Publications, 2001.

Ross-Larson, Bruce. *Edit Yourself: A Manual for Everyone Who Works With Words.* New York: W. W. Norton & Co., 1996.

Rude, Carolyn D. *Technical Editing.* 4th ed. New York: Longman, 2005.

Samson, Donald C., Jr. *Editing Technical Writing.* New York: Oxford University Press, 1993.

Stainton, Elsie Myers. *The Fine Art of Copyediting.* 2d ed. New York: Columbia University Press, 2001.

Tarutz, Judith A. *Technical Editing: The Practical Guide for Editors and Writers.* Cambridge, Mass.: Perseus Publishing, 1992.

Venolia, Jan. *Rewrite Right! Your Guide to Perfectly Polished Prose.* 2d ed. Berkeley, Calif.: Ten Speed Press, 2000.

Venolia, Jan. *Write Right! A Desktop Digest of Punctuation, Grammar, and Style.* 4th ed. Berkeley, Calif.: Ten Speed Press, 2001.

Graphics and Illustration

Bertoline, Gary R., and Eric Wiebe. *Technical Graphics Communication.* 4th ed. New York: McGraw-Hill Science/Engineering/Math, 2008.

Deemer, Joe. *Glossary of Graphic Communications.* 4th ed. Sewickley, Penn.: PIA/GATF Press, 2008.

Dreyfus, Henry. *Symbol Sourcebook: An Authoritative Guide to International Graphic Symbols.* New York: John Wiley & Sons, 1984.

Evans, Poppy. *Designer's Survival Manual: The Insider's Guide to Working With Illustrators, Photographers, Printers, Web Engineers, and More.* Cincinnati, Ohio: North Light Books, 2001.

Horton, William. *The Icon Book: Visual Symbols for Computer Systems and Documentation.* New York: John Wiley & Sons, 1994.

Tufte, Edward R. *Envisioning Information.* Cheshire, Conn.: Graphics Press, 1990.

Tufte, Edward R. *The Visual Display of Quantitative Information.* 2d ed. Cheshire, Conn.: Graphics Press, 2001.

Tufte, Edward R. *Visual Explanations: Images and Quantities, Evidence and Narrative.* Cheshire, Conn.: Graphics Press, 1997.

Ware, Colin. *Information Visualization: Optimizing Design for Human Perception.* 2d ed. San Francisco: Morgan Kaufmann, 2004.

Williamson, Hugh Albert Fordyce. *Methods of Book Design: The Practice of an Industrial Craft.* 3d ed. New Haven, Conn.: Yale University Press, 1983.

HTML

Burns, Joe. HTML Goodies web site at `http://www.htmlgoodies.com/`.

Graham, Ian S. *The HTML 4.0 Sourcebook: A Complete Guide to HTML 4.0.* New York: John Wiley & Sons, 1998.

Morris, Mary E.S., and John E. Simpson. *HTML for Fun and Profit.* 3d ed. Palo Alto, Calif.: Sun Microsystems Press, 1998.

Musciano, Chuck, and Bill Kennedy. *HTML & XHTML: The Definitive Guide.* 6th ed. Sebastopol, Calif.: O'Reilly & Associates, 2006.

Willard, Wendy. *HTML: A Beginner's Guide.* 4th ed. Emeryville, Calif.: McGraw-Hill/Osborne Media, 2009.

Indexing

Ament, Kurt. *Indexing: A Nuts-and-Bolts Guide for Technical Writers.* Norwich, N.Y.: William Andrew Publishing, 2001.

The American Society of Indexers web site at `http://asindexing.org`.

Bonura, Larry. *The Art of Indexing.* New York: John Wiley & Sons, 1994.

Brenner, Diane, and Marilyn Rowland, eds. *Beyond Book Indexing: How to Get Started in Web Indexing, Embedded Indexing, and Other Computer-Based Media.* Medford, N.J.: Information Today, 2000.

Lathrop, Lori. *An Indexer's Guide to the Internet.* 2d ed. Medford, N.J.: Information Today, 1995.

Mulvany, Nancy C. *Indexing Books*. 2d ed. Chicago: University of Chicago Press, 2005.

Wellisch, Hans H. *Indexing From A to Z*. 2d ed. New York: H. W. Wilson, 1995.

Information Mapping

Horn, Robert E. *Mapping Hypertext: The Analysis, Organization, and Display of Knowledge for the Next Generation of On-line Text and Graphics*. Lexington, Mass.: Lexington Institute, 1990.

The Information Mapping Method. Available at `http://www.infomap.com`.

Trubiano, John, and Gerard W. Paradis. *Demystifying ISO 9001:2000: Information Mapping's Guide to the ISO 9001 Standard, 2000 Version*. 2d ed. Upper Saddle River, N.J.: Prentice Hall, 2001.

Wycoff, Joyce. *Mindmapping: Your Personal Guide to Exploring Creativity and Problem-Solving*. New York: Berkley Books, 1991.

Internationalization and Localization

Bosley, Deborah S. *Global Contexts: Case Studies in International Technical Communication*. New York: Longman, 2000.

del Galdo, Elisa M., and Jakob Nielsen. *International User Interfaces*. New York: John Wiley & Sons, 1996.

Esselink, Bert. *A Practical Guide to Localization*. Philadelphia: John Benjamins Publishing, 2000.

Kohl, John R. *The Global English Style Guide: Writing Clear, Translatable Documentation for a Global Market*. Cary, N.C.: SAS Press, 2008.

Nielsen, Jakob. *Designing User Interfaces for International Use*. Amsterdam: Elsevier/North Holland, 1990.

Savourel, Yves. *XML Internationalization and Localization*. Indianapolis, Ind.: Sams Publishing, 2001.

Tuthill, Bill, and David Smallberg. *Creating Worldwide Software: Solaris International Developer's Guide*. 2d ed. Palo Alto, Calif.: Sun Microsystems Press, 1997.

Yunker, John. *Beyond Borders: Web Globalization Strategies*. Indianapolis, Ind.: New Riders Publishing, 2002.

Legal Issues

Bouchoux, Deborah E. *Protecting Your Company's Intellectual Property: A Practical Guide to Trademarks, Copyrights, Patents, and Trade Secrets.* New York: AMACOM, 2001.

Elias, Stephen. *Patent, Copyright & Trademark: A Desk Reference to Intellectual Property Law.* 10th ed. Berkeley, Calif.: Nolo Press, 2009.

Strong, William S. *The Copyright Book: A Practical Guide*. 5th ed. Cambridge, Mass.: MIT Press, 1999.

United States Copyright Office web site at `http://www.loc.gov/copyright/`.

United States Patent and Trademark Office web site at `http://www.uspto.gov`.

Online Help

Hedtke, John, and Brenda Huettner. *RoboHelp for the Web*. Includes CD-ROM. Plano, Texas: Wordware Publishing, 2002.

Help Technology Centre: Resources and Techniques for Help Systems. Available at `http://mvps.org/htmlhelpcenter`.

Klein, Jeannine M. E. *Building Enhanced HTML Help With DHTML and CSS.* Upper Saddle River, N.J.: Prentice Hall, 2000.

Weber, Jean Hollis. *Is the Help Helpful? How to Create Online Help That Meets Your Users' Needs*. Milwaukee, WI: Hentzenwerke, 2004.

Wickham, Daina Pupons, Debra L. Mayhew, Teresa Stoll [et al.]. *Designing Effective Wizards: A Multidisciplinary Approach.* Includes CD-ROM. Upper Saddle River, N.J.: Prentice Hall, 2001.

Online Writing Style

Gahran, Amy. *Contentious: The Web-zine for Writers, Editors, and Others Who Create Content for Online Media*. Available at `http://www.contentious.com`. 1998–2002.

Garrand, Timothy. *Writing for Multimedia and the Web: A Practical Guide to Content Development for Interactive Media*. 3d ed. Oxford, U.K.: Elsevier/Focal Press, 2006.

Hammerich, Irene, and Claire Harrison. *Developing Online Content: The Principles of Writing and Editing for the Web*. New York: John Wiley & Sons, 2001.

Horton, William. *Designing and Writing Online Documentation: Hypermedia for Self-Supporting Products*. 2d ed. New York: John Wiley & Sons, 1994.

Kilian, Crawford. *Writing for the Web (Self-Counsel Writing Series)*. 3d ed. Bellingham, Wash.: Self-Counsel Press, 2007.

Lynton, Jennifer, and Kylene Bruski. *Introduction to DITA: A User Guide to the Darwin Information Typing Architecture*. Denver: Comtech, 2006.

McGovern, Gerry, Rob Norton, and Catherine O'Dowd. *The Web Content Style Guide: An Essential Reference for Online Writers, Editors, and Managers*. Upper Saddle River, N.J.: Prentice Hall, 2001.

Nielsen, Jakob. "Writing for the Web." Useit.com. Available at `http://www.useit.com/papers/webwriting`.

Pfaffenberger, Bryan. *The Elements of Hypertext Style*. Boston: AP Professional, 1997.

Price, Jonathan, and Lisa Price. *Hot Text: Web Writing That Works*. Indianapolis, Ind.: New Riders Publishing, 2002.

Redish, Janice C. *Letting Go of the Words: Writing Web Content That Works*. San Francisco: Morgan Kaufmann, 2007.

Troffer, Alysson. "How to Write Effectively Online." Available at `http://homepage.mac.com/alysson/webfolio.html`.

Walker, Janice R., and Todd Taylor. *The Columbia Guide to Online Style*. 2d ed. New York: Columbia University Press, 2006.

Platform Style Guides

Apple Computer, Inc. *Macintosh Human Interface Guidelines*. Reading, Mass.: Addison-Wesley, 1992.

CDE Documentation Group. *Common Desktop Environment 1.0 Programmer's Guide*. Reading, Mass.: Addison-Wesley, 1995.

Commodore-Amiga, Inc. *Amiga User Interface Style Guide*. Reading, Mass.: Addison-Wesley, 1991.

Fountain, Anthony, and Paula Ferguson. *Motif Reference Manual: For Motif 2.1, Vol. 6*. 2d ed. Sebastopol, Calif.: O'Reilly & Associates, 2000.

GNOME Documentation Project. *GNOME Documentation Style Guide*, 2003.

Microsoft Corporation. *The Windows Interface: An Application Design Guide*. 2d ed. Redmond, Wash.: Microsoft Press, 1995.

Open Software Foundation. *OSF/Motif Style Guide: Revision 1.2 (for OSF/Motif Release 1.2)*. Upper Saddle River, N.J.: Prentice Hall PTR, 1993.

Sun Microsystems, Inc. *Java Look and Feel Design Guidelines*. 2d ed. Reading, Mass.: Addison-Wesley, 2001.

Printing

Adams, J. Michael, and Penny Ann Dolin. *Printing Technology*. 5th ed. Clifton, N.Y.: Delmar Learning, 2001.

Beach, Mark, and Eric Kenly. *Getting It Printed: How to Work With Printers and Graphic Imaging Services to Assure Quality, Stay on Schedule and Control Costs*. 4th ed. Cincinnati, Ohio: North Light Books, 2004.

O'Quinn, Donnie. *Print Publishing: A Hayden Shop Manual*. 2d ed. Indianapolis, Ind.: Que, 2000.

Project Management

Brooks, Frederick P. *The Mythical Man-Month: Essays on Software Engineering*. 20th anniversary ed. Reading, Mass.: Addison-Wesley, 1995.

DeMarco, Tom. *Controlling Software Projects: Management, Measurement, Measurement & Estimation*. Upper Saddle River, N.J.: Prentice Hall PTR, 1998.

Faulconbridge, R. Ian, and Michael J. Ryan. *Managing Complex Technical Projects: A Systems Engineering Approach*. Norwood, Mass.: 2002.

Hackos, JoAnn T. *Managing Your Documentation Projects*. New York: John Wiley & Sons, 1994.

Humphrey, Watts S. *Introduction to the Team Software Process*. Reading, Mass.: Addison-Wesley, 1999.

Humphrey, Watts S. *Managing the Software Process*. Reading, Mass.: Addison-Wesley, 1989.

Kerzner, Harold. *Project Management: A Systems Approach to Planning, Scheduling, and Controlling*. 9th ed. New York: John Wiley & Sons, 2005.

Lewis, James P. *Project Planning, Scheduling & Control*. 4th ed. New York: McGraw-Hill Trade, 2005.

McConnell, Steve C. *Software Project Survival Guide*. Redmond, Wash.: Microsoft Press, 1998.

Murch, Richard. *Project Management: Best Practices for IT Professionals*. Upper Saddle River, N.J.: Prentice Hall PTR, 2000.

Nicholas, John M., and Herman Steyn. *Project Management for Business and Technology: Principles and Practice*. Harlow, Essex, U.K.: Elsevier/Butterworth-Heinemann, 2008.

Phillips, Joseph. *IT Project Management: On Track From Start to Finish*. Includes CD-ROM. 2d ed. Emeryville, Calif.: McGraw-Hill/Osborne Media, 2004.

Reference Works

The American Heritage Dictionary of the English Language. 4th ed. Boston: Houghton Mifflin Company, 1996.

The Chicago Manual of Style. 15th ed. Chicago: University of Chicago Press, 2003.

The Chicago Manual of Style FAQ. Available at `http://www.press.uchicago.edu/Misc/Chicago/cmosfaq/cmosfaq.html`.

ComputerUser High-Tech Dictionary. Available at `http://www.computeruser.com/resources/dictionary`.

FOLDOC: Free On-Line Dictionary of Computing. Available at `http://foldoc.org`.

The Gregg Reference Manual. 9th ed. Woodland Hills, Calif.: Glencoe/McGraw-Hill, 2001.

IEEE 100: The Authoritative Dictionary of IEEE Standards Terms. 7th ed. New York: Institute of Electrical and Electronics Engineers, Inc., 2001.

McGraw-Hill Dictionary of Scientific and Technical Terms. 6th ed. New York: McGraw-Hill, 2002.

Merriam-Webster's Collegiate Dictionary. 11th ed. Springfield, Mass.: Merriam-Webster, 2003.

Merriam-Webster's Collegiate Thesaurus. Springfield, Mass.: Merriam-Webster, 1994.

Merriam-Webster's Medical Desk Dictionary. 2d ed. Clifton, N.Y.: Delmar Learning, 2002.

Merriam-Webster Online. Available at `http://www.m-w.com`.

Microsoft Corporation. *The Microsoft Manual of Style for Technical Publications*. 3d ed. Redmond, Wash.: Microsoft Press, 2003.

Microsoft Corporation. *Microsoft Press Computer Dictionary*. 5th ed. Redmond, Wash.: Microsoft Press, 2002.

The New Hacker's Dictionary. 3d ed. Cambridge, Mass.: MIT Press, 1996.

The New York Public Library Writer's Guide to Style and Usage. 1st ed. New York: Harper Collins, 1994.

OneLook Dictionary Search. Available at `http://www.onelook.com`.

Publications Manual of the American Psychological Association. 5th ed. Washington, D.C.: American Psychological Association, 2001.

Random House Webster's College Dictionary. New York: Random House, 2005.

Roget's II: The New Thesaurus. 3d ed. Boston: Houghton Mifflin Company, 2003.

The Synonym Finder. Completely revised by J. I. Rodale, Laurence Urdang, and Nancy LaRoche, eds. [et al.] New York: Time-Warner Books, 1986.

Webster's New World Computer Dictionary. 10th ed. New York: John Wiley & Sons, 2003.

Webster's Third New International Dictionary. Includes CD-ROM. Springfield, Mass.: Merriam-Webster, 2000.

Whatis?com: Definitions for Thousands of the Most Current IT-Related Words. Available at `http://whatis.techtarget.com`.

Words into Type. 3d ed. Upper Saddle River, N.J.: Prentice Hall, 1974.

SGML and XML

Bryan, Martin. *SGML: An Author's Guide to the Standard Generalized Markup Language.* Wokingham, England; Reading, Mass.: Addison-Wesley, 1988.

Bryan, Martin. *SGML and HTML Explained.* 2d ed. Harlow, England; Reading, Mass.: Addison-Wesley Longman, 1997.

Eckstein, Robert. *XML Pocket Reference.* 2d ed. Sebastopol, Calif.: O'Reilly & Associates, 2001.

Goldfarb, Charles F., with Yuri Rubinsky. *The SGML Handbook.* Oxford: Clarendon Press; Oxford and New York: Oxford University Press, 1990.

Harold, Elliotte Rusty, and W. Scott Means. *XML in a Nutshell.* 3d ed. Sebastopol, Calif.: O'Reilly & Associates, 2004.

Maler, Eve, with Jeanne El Andaloussi. *Developing SGML DTDs: From Text to Model to Markup.* Upper Saddle River, N.J.: Prentice Hall PTR, 1995.

OASIS. Cover Pages standards web site at `http://xml.coverpages.org/sgml.html`.

Pfaffenberger, Bryan. *Web Publishing With XML in Six Easy Steps.* Boston: AP Professional, 1998.

Ray, Erik T. *Learning XML.* 2d ed. Sebastopol, Calif.: O'Reilly & Associates, 2003.

Van Herwijnen, Eric. *Practical SGML.* 2d ed. Boston: Kluwer Academic Publishers, 1994.

Walsh, Norman, and Leonard Muellner. *DocBook: The Definitive Guide.* Sebastopol, Calif.: O'Reilly & Associates, 1999.

Typography

Bringhurst, Robert. *Elements of Typographic Style*. 3d ed. Point Roberts, Wash.: Hartley & Marks, 2004.

Campbell, Alastair. *The Designer's Lexicon: The Illustrated Dictionary of Design, Printing, and Computer Terms*. San Francisco: Chronicle Books, 2000.

Craig, James, and William Bevington. Susan E. Meyer, ed. *Designing With Type: A Basic Course in Typography*. 5th ed. New York: Watson-Guptill Publications, Inc., 2008.

Felici, James. *The Complete Manual of Typography*. Berkeley, Calif.: Peachpit Press, 2003.

Spiekermann, Erik, and E. M. Ginger. *Stop Stealing Sheep & Find Out How Type Works*. 2d ed. Mountain View, Calif.: Adobe Press, 2002.

Williams, Robin. *The Non-Designer's Design Book: Design and Typographic Principles for the Visual Novice*. Berkeley, Calif.: Peachpit Press, 1994.

Usability Testing

Barnum, Carol M. *Usability Testing and Research*. Reading, Mass.: Addison-Wesley Longman, 2001.

Dumas, Joseph S., and Janice C. Redish. *A Practical Guide to Usability Testing*. Bristol, U.K.: Intellect, 1999.

Rubin, Jeffrey, and Dana Chisnell. *Handbook of Usability Testing: How to Plan, Design, and Conduct Effective Tests*. 2d ed. New York: John Wiley & Sons, 2008.

User Interfaces

Beyer, Hugh, and Karen Holtzblatt. *Contextual Design: A Customer-Centered Approach to Systems Designs*. San Francisco: Morgan Kaufmann, 1998.

Coe, Marlana. *Human Factors for Technical Communicators*. New York: John Wiley & Sons, 1996.

Cooper, Alan. *The Inmates Are Running the Asylum: Why High Tech Products Drive Us Crazy and How to Restore the Sanity*. Indianapolis, Ind.: Sams Publishing, 1999.

Cooper, Alan, and Robert M. Reimann. *About Face 3: The Essentials of User Interface Design*. 3d ed. New York: John Wiley & Sons, 2007.

Galitz, Wilbert O. *The Essential Guide to User Interface Design: An Introduction to GUI Design Principles and Techniques*. 2d ed. New York: John Wiley & Sons, 2002.

Hackos, JoAnn T., and Janice C. Redish. *User and Task Analysis for Interface Design.* New York: John Wiley & Sons, 1998.

Hix, Deborah, and H. Rex Hartson. *Developing User Interfaces: Ensuring Usability Through Product & Process.* New York: John Wiley & Sons, 1993.

Isaacs, Ellen, and Alan Walendowski. *Designing from Both Sides of the Screen: How Designers and Engineers Can Collaborate to Build Cooperative Technology.* Indianapolis, Ind.: New Riders Publishing, 2001.

Johnson, Jeff. *GUI Bloopers 2.0: User-Interface Don'ts and Do's for Software Developers and Managers.* 2d ed. San Francisco: Morgan Kaufmann, 2007.

Laurel, Brenda, ed. *The Art of Human-Computer Interface Design.* Reading, Mass.: Addison-Wesley, 1990.

Laurel, Brenda. *Computers as Theatre.* Reading, Mass.: Addison-Wesley, 1993.

Mayhew, Deborah J. *Principles and Guidelines in Software User Interface Design.* Upper Saddle River, N.J.: Prentice Hall, 1997.

Mayhew, Deborah J. *The Usability Engineering Lifecycle: A Practitioner's Guide to User Interface Design.* San Francisco: Morgan Kaufman, 1999.

Nielsen, Jakob. *Usability Engineering.* San Francisco: Morgan Kaufmann, 1994.

Nielsen, Jakob. useit.com: Jakob Nielsen's site (Usability and Web Design). Available at `http://useit.com`.

Norman, Donald A. *The Design of Everyday Things.* New York: Basic Books, 2002.

Perlman, Gary. Suggested Readings in Human-Computer Interaction (HCI), User Interface (UI) Development, & Human Factors (HF). Available at `http://www.hcibib.org/readings.html`. Gary Perlman, 1993–2001.

Raskin, Jef. *The Humane Interface: New Directions for Designing Interactive Systems.* Harlow, Essex, U.K.: Pearson Education, 2000.

Society for Technical Communication Usability SIG web site. Available at `http://www.stcsig.org/usability/resources/index.html`.

Tognazzini, Bruce. *Tog on Interface.* Reading, Mass.: Addison-Wesley, 1992.

Vredenburg, Karel, Scott Isensee, and Carol Righi. *User-Centered Design: An Integrated Approach.* Harlow, Essex, U.K.: Pearson Education, 2001.

Web and Internet Publishing

Burdman, Jessica R. *Collaborative Web Development: Strategies and Best Practices for Web Teams.* Includes CD-ROM. Harlow, Essex, U.K.: Pearson Education, 2000.

Flanders, Vincent. *Son of Web Pages That Suck: Learn Good Design by Looking at Bad Design.* 2d ed. Includes CD-ROM. Alameda, Calif.: Sybex, 2002.

Garrett, Jesse James. *The Elements of User Experience: User-Centered Design for the Web.* Indianapolis, Ind.: New Riders Publishing, 2002.

Goto, Kelly, and Emily Cotler. *Web ReDesign 2.0: Workflow That Works.* 2d ed. Indianapolis, Ind.: New Riders Publishing, 2004.

Hackos, JoAnn T. *Content Management for Dynamic Web Delivery.* New York: John Wiley & Sons, 2002.

Hackos, JoAnn T., and Dawn M. Stevens. *Standards for Online Communication: Publishing Information for the Internet/World Wide Web/Help Systems/Corporate Intranets.* New York: John Wiley & Sons, 1997.

Krug, Steve. *Don't Make Me Think: A Common Sense Approach to Web Usability.* 3d ed. Indianapolis, Ind.: New Riders Publishing, 2005.

Lynch, Patrick J., and Sarah Horton. *Web Style Guide: Basic Design Principles for Creating Web Sites.* New Haven, Conn.: Yale University Press, 1999.

National Cancer Institute. *Research-Based Web Design and Usability Guidelines.* 2006 edition. Available at `http://usability.gov/guidelines`.

Nielsen, Jakob. *Designing Web Usability: The Practice of Simplicity.* Indianapolis, Ind.: New Riders Publishing, 2000.

Nielsen, Jakob. *Multimedia and Hypertext: The Internet and Beyond.* Boston: AP Professional, 1995.

Nielsen, Jakob, and Marie Tahir. *Homepage Usability: 50 Websites Deconstructed.* Indianapolis, Ind.: New Riders Publishing, 2001.

Rosenfeld, Louis, and Peter Morville. *Information Architecture for the World Wide Web: Designing Large-Scale Web Sites.* 3d ed. Sebastopol, Calif.: O'Reilly & Associates, 2009.

Slatin, John M., and Sharron Rush. *Maximum Accessibility: Making Your Web Site More Usable for Everyone.* Reading, Mass.: Addison Wesley Professional, 2002.

Williams, Robin, and John Tollett. *The Non-Designer's Web Book: An Easy Guide to Creating, Designing, and Posting Your Own Web Site.* 3d ed. Berkeley, Calif.: Peachpit Press, 2005.

Wodtke, Christina. *Information Architecture: Blueprints for the Web.* 2d ed. Indianapolis, Ind.: New Riders Publishing, 2009.

Wikis, Blogs, and Social Media

Bly, Robert W. *Blog Schmog: The Truth About What Blogs Can (and Can't) Do for Your Business.* Nashville, Tenn.: Thomas Nelson, 2007.

Coleman, David. *42 Rules for Successful Collaboration: A Practical Approach to Working With People, Processes, and Technology.* Cupertino, Calif.: Super Star Press, 2009.

Coleman, David, and Stewart Levine. *Collaboration 2.0: Technology and Best Practices for Successful Collaboration in a Web 2.0 World.* Cupertino, Calif.: Super Star Press, 2008.

Holtz, Shel, and Ted Demopoulos. *Blogging for Business: Everything You Need to Know and Why You Should Care.* New York: Kaplan Publishing, 2006.

Huettner, Brenda, Char James-Tanny, and Kit Brown. *Managing Virtual Teams: Getting the Most from Wikis, Blogs, and Other Collaborative Tools.* Sudbury, Mass.: WordWare Press, 2006.

Leuf, Bo, and Ward Cunningham. *The Wiki Way: Quick Collaboration on the Web.* Reading, Mass.: Addison-Wesley, 2001.

Li, Charlene, and Josh Bernoff. *Groundswell: Winning in a World Transformed by Social Technologies.* Boston: Harvard Business School Press, 2008.

Mader, Stewart. *Wikipatterns: A Practical Guide to Improving Productivity and Collaboration in Your Organization.* Indianapolis, Ind.: Wiley Publishing, 2008.

Safko, Lon, and David K. Brake. *The Social Media Bible: Tactics, Tools, and Strategies for Business Success.* Hoboken, N.J.: John Wiley & Sons, 2009.

Scoble, Robert, and Shel Israel. *Naked Conversations: How Blogs Are Changing the Way Businesses Talk With Customers.* Hoboken, N.J.: John Wiley & Sons, 2006.

Writing Standards

Barzun, Jacques. *Simple & Direct: A Rhetoric for Writers.* 4th ed. New York: Harper Collins, 2001.

Brooks, Brian S. *Working With Words: A Handbook for Writers and Editors.* 5th ed. New York: St. Martin's Press, 2002.

Dupre, Lyn. *Bugs in Writing: A Guide to Debugging Your Prose.* Rev. ed. Reading, Mass.: Addison-Wesley, 1998.

Ede, Lisa S. *Work in Progress: A Guide to Writing and Revising,* 4th ed. New York: St. Martin's Press, 1998.

Flesch, Rudolf, and A.H. Lass. *The Classic Guide to Better Writing.* 50th anniversary ed. New York: Harper Collins, 1996.

Strunk, William, Jr., and E. B. White. *The Elements of Style*. 50th anniversary ed. New York: Longman, 2008.

Williams, Joseph M. *Style: Ten Lessons in Clarity and Grace*. 8th ed. New York: Longman, 2005.

Young, Matt. *The Technical Writer's Handbook: Writing With Style and Clarity*. Mill Valley, Calif.: University Science Books, 2002.

Zinsser, William K. *On Writing Well*. 30th anniversary ed. New York: HarperCollins, 2006.

Writing Standards for Technical Writing

Alred, Gerald J., Charles T. Brusaw, and Walter E. Oliu. *Handbook of Technical Writing*. 9th ed. New York: St. Martin's Press, 2008.

Ament, Kurt. *Single Sourcing: Building Modular Documentation*. Park Ridge, N.J.: Noyes Publications, 2002.

Barker, Thomas T. *Writing Software Documentation: A Task-Oriented Approach*. 2d ed. Harlow, Essex, U.K.: Pearson Education, 2002.

Brogan, John A. *Clear Technical Writing*. New York: McGraw-Hill, 1973.

Burnett, Rebecca E. *Technical Communication*. 6th ed. Belmont, Calif.: Wadsworth Publishing, 2004.

Mager, Robert Frank. *Preparing Instructional Objectives*. 3d ed. Atlanta, Ga.: Center for Effective Performance, 1997.

Price, Jonathan. *How to Communicate Technical Information: A Handbook of Software and Hardware Documentation*. Redwood City, Calif.: Benjamin/Cummings Publishing Co., 1993.

Pringle, Alan S., and Sarah S. O'Keefe. *Technical Writing 101: A Real-World Guide to Planning and Writing Technical Documentation*. 2d ed. Cary, N.C.: Scriptorium Press, 2003.

Robinson, Patricia, and Ryn Etter. *Writing and Designing Manuals*. 3d ed. Boca Raton, Fla.: CRC Press, 2000.

Sides, Charles H. *How to Write & Present Technical Information*. 3d ed. Phoenix, Ariz.: Oryx Press, 1998.

Weiss, Edmond H. *How to Write Usable User Documentation*. 2d ed. Phoenix, Ariz.: Oryx Press, 1991.

Index

B

D

E

F

H

I

J

P

Q

R

S

T